Powerful Primary Geography

Powerful Primary Geography: A Toolkit for 21st-Century Learning explores the need for children to understand the modern world and their place in it. Dedicated to helping teachers inspire children's love of place, nature and geographical adventures through facilitating children's voice and developing their agency, this book explores the way playful opportunities can be created for children to learn how to think geographically, to solve real-life problems and to apply their learning in meaningful ways to the world around them.

Based on the very latest research, *Powerful Primary Geography* helps children understand change, conflict and contemporary issues influencing their current and future lives and covers topics such as:

- Weather and climate change
- Sustainability
- Engaging in their local and global community
- Graphicacy, map work and visual literacy
- Understanding geography through the arts.

Including several case studies from primary schools in Ireland, this book will help aid teachers, student teachers and education enthusiasts in preparing children for dealing with the complex nature of our contemporary world through artistic and thoughtful geography. Facilitating children's engagement as local, national and global citizens ensures geography can be taught in a powerful and meaningful manner.

Anne M. Dolan is a lecturer in primary geography with the Department of Learning, Society and Religious Education in Mary Immaculate College, University of Limerick, Ireland. Anne is passionate about primary geography and she is keen to share this passion with her students and primary teachers. She is particularly interested in creative approaches to geography, interdisciplinary collaboration and the use of the arts in geographical explorations. As a researcher, Anne focuses on outdoor learning, geoliteracy, children's concepts of place and creative approaches to geography.

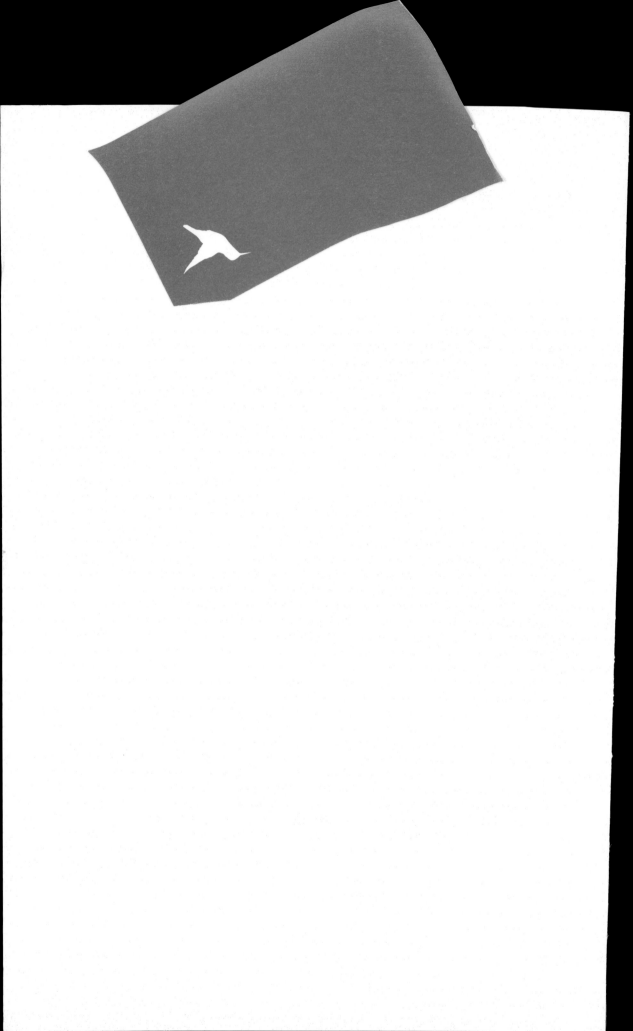

Powerful Primary Geography
A Toolkit for 21st-Century Learning

Anne M. Dolan

LONDON AND NEW YORK

First published 2020
by Routledge
2 Park Square, Milton Park, Abingdon, Oxon OX14 4RN

and by Routledge
52 Vanderbilt Avenue, New York, NY 10017

Routledge is an imprint of the Taylor & Francis Group, an informa business

© 2020 Anne M. Dolan

The right of Anne M. Dolan to be identified as author of this work has been asserted by them in accordance with sections 77 and 78 of the Copyright, Designs and Patents Act 1988.

All rights reserved. No part of this book may be reprinted or reproduced or utilised in any form or by any electronic, mechanical, or other means, now known or hereafter invented, including photocopying and recording, or in any information storage or retrieval system, without permission in writing from the publishers.

Trademark notice: Product or corporate names may be trademarks or registered trademarks, and are used only for identification and explanation without intent to infringe.

British Library Cataloguing-in-Publication Data
A catalogue record for this book is available from the British Library

Library of Congress Cataloging-in-Publication Data
A catalog record has been requested for this book

ISBN: 978-1-138-22650-0 (hbk)
ISBN: 978-1-138-22651-7 (pbk)
ISBN: 978-1-315-39754-2 (ebk)

Typeset in Bembo
by Swales & Willis, Exeter, Devon, UK

Contents

	Introduction	1
1	Powerful primary geography: Setting the scene	10
2	Powerful geographical thinking: Initiating investigations and enquiry-based learning	34
3	Teaching powerful geography through place	58
4	Playful approaches to powerful geography: Games, artefacts and fun	84
5	Teaching powerful geography through topics: Weather and climate change	113
6	Teaching powerful geography through graphicacy, map work and visual literacy	147
7	Teaching powerful geography through the arts	181
8	Powerful geography: Teaching citizenship, global learning and the Sustainable Development Goals	212
	Appendix 1: Card-sorting activity for teaching about volcanoes	246
	Appendix 2: Weather glossary	250
	Index	254

Introduction

In the 21st century, teachers and learners require curiosity, imagination, creativity, problem-solving skills, flexibility, digital learning and collaboration. The year 2080 will represent the beginning of retirement for children entering primary education today, all things being equal. Our children's journey towards 2080 will continue to be defined by changes in technology and globalisation. Change prompts new ideas and new solutions. Our children will have to be more knowledgeable, creative, innovative and flexible. They will have to be more resilient cognitively, socially and emotionally. Such resilience requires comprehensive education including environmental and international knowledge, 'the twin pillars of modern geography' (Bonnett, 2008: 54). International and environmental knowledge are absolutely critical in a globalised world defined by issues of injustice, inequality, overconsumption, political instability and climate change. Equally important is a pedagogy of hope whereby children can appreciate their own agency and role as problem-solvers.

While the emphasis in our schools today is on numeracy and maths, the demand for creative and performative skills from the economic, social, political and environmental sectors is as great as, if not greater than, the need for a good memory and factual knowledge. Children should develop informed opinions about local and global geographical issues. In order for this to happen they need to understand the basic physical and human processes involved. If geography, including a sound knowledge about people and places, is taught well children can develop the skills and ability to understand themselves, their place, their environment and other people and cultures, which collectively counts as powerful knowledge. In order to teach geography powerfully, teachers must have engaged previously with the subject in a robust manner.

Powerful primary geography helps children to understand the world around them, including people, systems, places, interactions and decisions which have an impact on their lives. It makes abstract geographical concepts accessible for children. Powerful primary geography is essentially about helping children, student teachers and teachers become excited about their world and their place in this world. For children and young people to participate effectively in this changing world, they must understand it.

This book shines a new lens on primary geography. It argues that geography can be taught in a stimulating fashion which facilitates children's voice and agency. The centrality of geography as part of the life cycle of the school is highlighted. In Bonnett's (2008: 6) words, this book illustrates how geography can 'find and impose order on a seemingly chaotic world'. Examples of empowering and inspirational geography are presented. The book is informed by the most recent research in geographical and pedagogical approaches to teaching and learning. Case studies for integrating geography teaching with other curricular areas including literacy, drama and art are provided.

Powerful geography embraces local studies, place-based learning and outdoor learning. Children have many outdoor experiences, albeit mixed in quality and quantity. They experience the world through journeys and interactions with their environment. Children negotiate and interact with a variety of landscapes (human and natural) on a daily basis. For example, children travel a variety of routes, they make decisions about places to visit and increasingly they use mobile hand-held devices (including mobile phones) to map their location and to communicate spatial information. More importantly, they have many geographical questions about the outdoors. Through these daily interactions and decisions they build up a wide knowledge base about the world, near and far, through a range of direct and indirect experiences. Formal geographical education, on the other hand, tends to take place indoors. Powerful primary geography prioritises outdoor learning in terms of fieldwork, using the locality for enquiry-based learning and experiential learning.

Powerful geography involves making connections. The global dimension is so much more than learning about people and places in other countries. It is about exploring interconnections between people and places. It is about exploring similarities and differences. Ultimately it is about considering our role as global citizens and the corresponding responsibilities which this entails. The impact of online interconnections brings the importance of geographical understanding to the fore. Harvesting data is now big business. Facebook's collection of data makes it one of the most influential organisations in the world. Our online activity, such as a Facebook like or a Google search, is constantly tracked. These interactions feed complex algorithms illustrating personal interests and purchasing preferences. This information is then used to target us as consumers with terrifying accuracy. Children need to be aware of this complex web which is shaping our lived experiences. Critical thinking skills are required to understand how opinions are shaped and formed by media and online influences. Powerful primary geography equips children to think critically and creatively about their world. It enables them to make connections between themselves and other people, other places, their environment and other ideas. It provides lifelong supports for consequential engagement in society. Placing the child at the centre of geographical encounters is essential for developmental and meaningful conceptual development (Figure 0.1).

This book is written for primary educators interested in learning about how to teach geography powerfully. The book is written to demonstrate that geography teaching and

Figure 0.1 Agency of the child is prioritised by powerful primary geography

Figure 0.2 Geography's big ideas

learning should be enjoyable, creative and empowering experiences for everyone – after all it is about the real world we live in! Geography is about places, people and issues, both local and global, and those should be of interest and concern to everyone. We are educating young citizens for a very different and rapidly changing world and geography allows us to teach what will be both useful and important to them now and in their future.

The book is evidence based as the material has emerged from the author's research work in schools and through consultation with teachers and children. There is a very strong commitment in the book to highlighting the voice of children, student teachers and teachers. This book is all about helping teachers to view the world through a geographical lens to develop their geographical imaginations (Martin, 2006) and to use geographical enquiry-based approaches in the classroom. The key concepts of place, space and environment provide an overarching framework for other substantive geographical concepts, as set out in Figure 0.2.

Children need to be able to make and understand connections within their world at local, national and international levels in an age-appropriate way. Moreover, children are superb at making these links and their insight often surprises teachers.

An ability to interpret diverse situations is also important. Helping children to become decision-makers helps them to appreciate the different sides of an argument or a proposal. Involving children in school-based decision-making can lay the foundation for geographical skill development. In circumstances where schools are planning an extension, a new garden, playground or a building asking children to research and pitch their ideas is a valuable learning opportunity for both children and the wider school community.

Practically everything has a geographical dimension (Dorling and Lee, 2016). Travelling involves spatial awareness and negotiation of local areas. Everyday events such as shopping and travelling in the local area all have geographical elements. Shopping involves purchasing goods which have come from many locations. Finding out the origin and journey of favourite foods is an important geographical exercise.

Part of making links and connections involves intercultural education. We live in a multicultural society; children need cultural sensitivity and the ability to be empathetic. Learning how to be culturally sensitive is a long-term process, which needs to be prioritised by school management and addressed by teachers and children. Children need to understand the relationships which exist between cultures, including differences and similarities. A child should understand that another child in a different part of the world may not speak the same language, practise the same religion or attend the same type of school, yet both are connected through the common experiences of childhood. Both like to play games, enjoy learning and participate in their own local cultural events. International events such as the Olympics and the rugby and soccer World Cups, along with local sports events, bring out a range of flags as we reaffirm our local and national identities. These events also provide opportunities for deeper geographical enquiry and investigation; for example, children can investigate the advantages and disadvantages of one country's bid to host an international sports event.

Memories of learning geography in school

Primary education geographers were asked about their formative experiences in geography (Catling et al., 2010). In this study several significant features emerged, including experiences of 'freedom to roam' locally, family holidays, outings and trips abroad, access to maps, fieldwork activities at primary or secondary school and with other organisations, and the impact of a good teacher. Yet, memories of geography for some student teachers tend to be limited to tracing maps, rote learning of rivers and mountains, as well as boring encounters with textbooks (Waldron et al., 2009; Dolan et al., 2014). Those with positive memories remember the teacher who inspired them, the teacher who engaged the children in enquiry projects and the teacher who brought the children outside regardless of weather conditions. Positive memories of geographical learning are closely associated with experiential learning outside (ibid.).

As part of the research for this book, I asked student teachers to reflect upon their memories of learning geography and here are some examples of positive memories. These memories speak for themselves.

> Our teacher brought us on walks every month. We collected leaves and cones, we became familiar with our area and back in the classroom we recorded these walks through pictures, writings and project work.
>
> Every September our teacher talked about his trips during the summer. He brought in a range of pictures, photographs and artefacts. We generally did a massive project on a particular country which was supplemented by the teacher's stories, by our own research and by an occasional guest speaker. It was the best time of the year.
>
> I had an amazing teacher when it came to geography. He was a massive fan of the Burren and we studied its landscape a lot and also visited there. He made Geography engaging, creative and fun. We investigated stuff ourselves and found out things for ourselves. It was thrilling discovering new stuff for ourselves.

Many student teachers vividly recalled rote learning of lists of geographical features. Several included these as negative memories. However, a minority of students claimed that it was important to know the towns and cities of Ireland as well as other factual information. Student teachers recalled a series of lists of place names for towns, cities, rivers, mountains and coastal features. In some cases rhymes or mnemonics were used to assist the process, e.g. FAT DAD as an acronym for the counties of Northern Ireland namely: Fermanagh, Antrim, Tyrone, Derry, Armagh and Down. Rhymes to aid memorisation were also noted. One student could sing the 32 counties on the island of Ireland. Often referred to as 'the capes and bays' approach to learning geography, this involves the memorisation of long lists of capes, bays, rivers, mountains, islands and products from different countries.

Also associated with pub quizzes and the game Trivial Pursuit, this kind of knowledge is one dimension of geography, a factual dimension where answers can be obtained from a Google search, globe, atlas or a map. This is however only one dimension of geography.

Twenty-first-century competencies and skills

We live in a rapidly changing world. Simply reprocessing knowledge will not be sufficient to address the challenges facing young people in the future. Teachers have to prepare children for jobs which have not yet been created, technologies which have yet to be invented and a range of unknown opportunities and challenges which have yet to become apparent. Some of our traditional concepts of the world are no longer relevant. For instance, the traditional dominance of America and Europe in global trade is no longer guaranteed due to the rapid growth of the Chinese and Indian economies.

The biggest increase in the number of skilled individuals is not in the West but in those regions where the population growth is greatest – in Asia, Africa and South America. Delhi's population, for example, currently exceeds 26 million people and is predicted to rise exponentially in the future. Hence, an Indian engineer can do the same job as his or her British or Irish counterpart but for a quarter of the price. In Kenya, the M-Pesa mobile money system allows Kenyans to transfer money by text. It does not require a bank account or an iPhone; all it requires is a 20-year-old Nokia. M-Pesa has allowed Kenya to leapfrog the traditional 20th-century stages of development – infrastructure and banks – with the most basic technology available. A gateway to digital consumerism, M-Pesa has disrupted the traditional function of money by colonising its functions, including purchase of goods, loans, wage payments and even online gambling (Peretti, 2017).

However, the narrative of change is only partially true. Biesta (2015) challenges the universal claim that all aspects of life are changing, illustrating it as half-true and half-false. While change is unprecedented in economic matters and access to information, there are people in parts of the world for whom very little has changed. For instance, many families are struggling to find clean water, feed their families and earn an income. For these families the mantra of universal change offers little hope. Biesta (2015: 6) suggests that the narrative of change and globalisation functions as an ideology in which 'half-truths mask as much as they express'. This is a powerful reminder for all of us in education to question some of the grand narratives which are taken for granted in school and society today.

Notwithstanding the importance of knowledge, including geographical knowledge, meeting the demands of today's world requires a shift in how we conceptualise teaching and learning. The adoption of 21st-century skills in our schools is widely recommended (Colvin and Edwards, 2018; Schleicher, 2018). This involves a movement from a measurement of knowledge to measuring children's ability to think creatively and critically, examine problems, gather information and make well-informed decisions using technology. Twenty-first-century learning involves a strong emphasis on problem-solving and decision-making and the ability to provide justification for the solution offered. Children as 21st-century learners need to be able to understand their world, their place in the world and how to interact with their world confidently and competently. Twenty-first-century learning is about the capacity to live in a dynamic world as an engaged citizen.

However, commentators such as Biesta caution against making this into a universal claim as it is accurate in some cases but not in its totality. He agrees that we need a broader conceptualisation of education, but it has to begin with questions of democracy, ecology and care as 'orientation points' (2015: 7). In light of the half-truths which dominate

6 Introduction

Table 0.1 Guide to comparative age of children in different geographical jurisdictions

Age	Ireland	Northern Ireland	California State, USA	Great Britain	
4–5	Junior Infants	Primary 1	P-K	Foundation Stage	Reception
5–6	Senior Infants	Primary 2	K		Year 1
6–7	1st Class	Primary 3	Grade 1	Key Stage 1	Year 2
7–8	2nd Class	Primary 4	Grade 2		Year 3
8–9	3rd Class	Primary 5	Grade 3	Key Stage 2	Year 4
9–10	4th Class	Primary 6	Grade 4		Year 5
10–11	5th Class	Primary 7	Grade 5		Year 6
11–13	6th Class	Year 8	Grade 6	Key Stage 3	Year 7

educational discourse, Biesta argues that we need to shift our focus from survival to meaningful living. Survival is about adapting to changing circumstances but perhaps we need to question the essential nature of these circumstances and their appropriateness for us and for society as a whole. Perhaps we need to envision a new set of circumstances which will ensure that we are 'more sustainable, more caring and more democratic' (2015: 7). In order to address these circumstances, Biesta returns to the basics of education and suggests that schools are places where children can practise living in a 'grown-up way' and where we can ask the question: What is desirable (a) for our own life (b) for others and (c) for the life we live on a vulnerable planet? While these are essentially geographical questions they also provide a valuable context for the development of 21st-century competencies. The development of 21st-century competencies should not in itself constitute the end goal but should instead provide a framework for sound education which facilitates the development of the individual's full potential while also promoting the development of a more sustainable, more caring and more democratic society.

This book is written for all those involved in primary geography. While the book refers to children in terms of their age group, Table 0.1 provides a guide for children's ages applied in education frameworks from different jurisdictions.

Overview of chapters

This book describes and showcases the concept of *powerful primary geography*. In light of an overloaded curriculum, the limited time available for geography teaching, a heavy reliance on textbooks and the political supremacy of literacy and numeracy, this book offers recommendations to teachers for teaching geography powerfully and meaningfully. Each chapter documents theoretical and practical aspects for teaching primary geography.

Chapter 1: Powerful primary geography: Setting the scene

This chapter sets the scene for the book by exploring the nature of powerful primary geography. It calls on teachers to reflect on their own personal geographies, their memories of learning geography in school and their current experiences of teaching geography. It explores the nature of geography in general and primary geography in particular in the context of globalisation, increased inequality and an uncertain future. Primary geography as a powerful discipline is discussed and situated in a broader social, political and economic context. Powerful primary geography is presented in terms of knowledge, aesthetic qualities, enquiry and skill development. The potential of primary geography for developing 21st-century competencies is highlighted.

Chapter 2: Powerful geographical thinking: Initiating investigations and enquiry-based learning

This chapter examines the importance of geographical thinking, which is promoted through children asking questions and conducting investigations or enquiries. The idea of children 'doing' geography as opposed to 'learning' geography is a useful metaphor for teachers. Thinking about child engagement and what the child will be doing at different stages of the lesson or geographical enquiry moves the focus from the teacher to the child. This chapter focuses on geographical investigations and the skills required for these investigations from the perspective of children and teachers. The enquiry process begins with an enquiry question. This big question may arise from a series of smaller questions posed by the children. Through a process of clarification and prioritisation a final big question is agreed with the whole class. This forms the basis of the enquiry. Once the question is selected, the class and the teacher decide how this question is going to be answered, what data is going to be collected and how it is going to be collected. Following data collection, the information is categorised, analysed and conclusions are drawn. Generally the class has an opportunity to share its findings with a wider audience. Various initiatives for promoting enquiries are discussed, such as use of class mascots, mysteries and a class newspaper.

Chapter 3: Teaching powerful geography through place

Place-based education is about generating a deep knowledge of a particular place so that children will eventually care about landscape, nature and people linked to a place. The term 'sense of place' is often used to emphasise that places are significant because they are the focus of personal feelings. In this chapter, I discuss the concepts of place, sense of place, place-based education, place attachment, place names, outdoor learning and critical pedagogy of place.

The chapter includes a number of methodological frameworks for exploring place including '8-way thinking', inspired by Gardner's theory of multiple intelligence (2011). This suggests that we have at least eight intelligences, namely, linguistic, musical, logical-mathematical, spatial, bodily kinesthetic, interpersonal, intrapersonal and naturalistic. The latest addition, naturalistic intelligence, is defined as the person's ability to recognise and classify his or her natural environment. In order to develop this intelligence, opportunities for outdoor learning are essential. Gardner claims that just as children are ready to master language at an early age, so too are they predisposed to explore the world of nature. Other methodologies for exploring place include story, drama and local trails.

Chapter 4: Playful approaches to powerful geography: Games, artefacts and fun

In this chapter, I explore the concept of playful approaches to teaching geography. The concept of geographical playfulness is discussed, along with different types of play and various structured and unstructured materials which can promote play. The value of using and handling artefacts is discussed. The chapter discusses opportunities for children to handle rocks and make volcanoes. Geography games such as Monopoly and playful initiatives such as Mission Explore are examined.

8   *Introduction*

Chapter 5: Teaching powerful geography through topics: Weather and climate change

In the context of challenges posed by climate change, this chapter draws attention to the significance of children's relationship with weather. In this chapter, I explore different strategies for teaching weather, seasonal changes and climate change. There is a need to look beyond the ways children learn 'about' the weather (where this is presented as something separate to our human selves), to more situated and intertwined ways of learning 'in' and 'with' weather. I also focus on teaching unusual weather conditions and natural disasters as a gateway to climate change education. An analysis of climate change education, climate change denial and climate justice education is provided. Case studies from schools illustrate innovative climate change education.

Chapter 6: Teaching powerful geography through graphicacy, map work and visual literacy

Graphicacy, a form of visual literacy, is the ability to understand and present information in a visual manner through media such as maps, graphs, diagrams and drawings. Most of the information we acquire as children and adults is communicated through visual means. Therefore it makes sense that the development of visual literacy, graphicacy and mapping skills is promoted as part of geographical education in primary schools. This chapter explores a range of strategies for promoting children's graphicacy such as sketching, working with images and working as photographers. The chapter also explores a range of mapping strategies such as journey sticks, mapping through children's literature and problem-solving with maps.

Chapter 7: Teaching powerful geography through the arts

This chapter, with its specific focus on visual and environmental art, is a response to the increasing environmental issues in our society today. Sometimes teachers and children feel disempowered in light of the scale and complexity of geographical issues. This chapter illustrates how children and student teachers respond to contemporary local and global issues through collaborative art pieces, eco art, landscape boxes, architecture, sculpture, scrapbooks and geographical art in the school grounds. Through cross-curricular links with art, the creative process of viewing the world through an alternate lens is promoted. Art helps us to review and renew our geographical understanding of local, national and global issues by making the usual unusual, reframing the ordinary, catching us off guard and provoking us. Much of the philosophy underpinning this chapter is informed by enquiry, constructivism and problem-based learning. Exploring geography with, in and through art encompasses a learning process that is inherently experiential and open-ended.

Chapter 8: Powerful geography: Teaching citizenship, global learning and the Sustainable Development Goals

Growing up in a globalised world requires new approaches to education which develop in children a global dimension, a futures perspective and 21st-century competencies. In this chapter, I discuss the implications of living in a globalised society and the importance of global competencies as a set of requirements for living in and participating in a dynamic rapidly changing world. This chapter highlights strategies for teaching global and justice issues through primary geography. Specifically the chapter presents innovative approaches to global learning, and opportunities presented by the Sustainable Development Goals (SDGs). The challenges facing our world are significant. We can no longer afford to look the other way. By adopting a powerful approach to teaching geography, teachers and children will realise a sense of hope, appreciate their own agency and transform the teaching of primary geography.

References

Biesta, G. (2015) The duty to resist: Redefining the basics for today's schools. *RoSE–Research on Steiner Education*, 6, 1–11.

Bonnett, A. (2008) *What is geography?* London: Sage.

Catling, S., Greenwood, R., Martin, F. and Owens, P. (2010) Formative experiences of primary geography educators. *International Research in Geographical and Environmental Education*, 19(4), 341–350.

Colvin, R.L. and Edwards, V. (2018) Teaching for global competence in a rapidly changing world. OECD Publishing. https://asiasociety.org/sites/default/files/inline-files/teaching-for-global-competence-in-a-rapidly-changing-world-edu.pdf

Dolan, A.M., Waldron, F., Pike, S. and Greenwood, R. (2014) Student teachers' reflections on prior experiences of learning geography. *International Research in Geographical and Environmental Education*, 23(4), 314–330.

Dorling, D. and Lee, C. (2016) *Geography: Ideas in profile*. London: Profile Books.

Gardner, H. (2011) *Frames of mind: The theory of multiple intelligences*. New York: Basic Books.

Martin, F. (2006) *Teaching geography in primary schools: Learning to live in the world*. Cambridge: Chris Kington Publishing.

Peretti, J. (2017) *Done: The secret deals that are changing our world*. London: Hodder & Stoughton.

Schleicher, A. (2018) Preparing our youth for an inclusive and sustainable world. *The OECD PISA global competence framework OECD*. www.oecd.org/education/Global-competency-for-an-inclusive-world.pdf

Waldron, F., Pike, S., Greenwood, R., Murphy, C., O'Connor, G., Dolan, A. and Kerr, C. (2009). *Becoming a teacher: Primary student teachers as learners and teachers of history, geography and science: An all-Ireland study*. Standing Conference on Teacher Education North and South (Scotens).

1 Powerful primary geography
Setting the scene

Children as 21st-century learners need to be able to understand their world and their place in it. Prerequisites for 21st-century teachers and learners include curiosity, imagination, creativity, problem-solving skills and flexibility. Twenty-first-century competencies help children learn how to interact with their world confidently and competently. An ability to work collaboratively along with a requirement to be digitally literate are required in the 21st-century classroom and the workforce. Primary geography has the potential to help children learn about their place and their world in a manner which develops 21st-century competencies.

This book describes and showcases the concept of *powerful primary geography*. Enabling children to understand the world around them includes an appreciation of the people, systems, places, interactions and decisions which impact upon their lives. This discipline helps us to understand change, conflict and important issues which have a bearing on our current and future lives.

This chapter sets out to:

- frame the scene for the book as a whole by exploring the nature of powerful primary geography;
- present primary geography as a powerful discipline and situate it in a broader social, political and economic context;
- examine the potential political power of primary geography;
- discuss powerful primary geography in terms of knowledge, aesthetic qualities, enquiry and skill development; and
- provide examples of powerful geography in primary classrooms.

Definition of geography

Geography is a dynamic, living, contemporary and exciting subject. Bonnett describes geography as 'one of humanity's big ideas', 'a fundamental fascination' and 'a core component of a good education' (2008: 1). According to Dorling and Lee (2016: 6) 'it is about what is where and where is what; and why and when, and who and how. It is about exploring places and spaces.'

In the Irish primary curriculum, geography is defined as 'the study of the earth, its inhabitants and the relationships between them in the context of place, space, and the environment' (Department of Education and Skills (DES)/NCCA, 1999: 6). Both the Geographical Association Manifesto (GA, 2009) and Lambert (2009) use the term 'living geography'. While 'living geography' is about current issues, it recognises the past and it is also futures oriented, whereby children are encouraged to think about their personal, collective and spatial futures. Its emphasis on sustainability generates a synthesis across the physical and human worlds. A focus on local geography is set in wider spatial and global contexts requiring children to focus on specific geographical issues through interlocking scales. A key dimension of 'living geography' involves children working as investigators of processes that bring change to environments – these can be grouped as environmental (or 'physical'), social, economic or political. Finally, the concept 'living geography' presents a dynamic interpretation of geography which promotes a critical interrogation of key geographical ideas to support children's conceptual development of place, space and the environment. This concrete interpretation of 'living geography' makes geography meaningful and relevant, and embraces children's personal geographies.

According to the GA (2011) there are three key organising concepts for geography, namely place, space and environment, all illustrated in Table 1.1. These concepts or 'big ideas' provide a geographical lens through which we can understand the world. They can be used to inform decision-making, problem-solving, organisation of data, and planning for geographical enquiry.

Table 1.1 Adapted from 'What should be taught in school geography?' The Geography National Curriculum GA Curriculum Proposals and Rationale

Place (places, territories and regions)	Space (patterns and links)	Environment (physical and human interaction)	Geographical enquiry (procedures and tools)
Knowledge of the local place in its community and regional context	Knowledge and understanding of economic patterns of production, distribution and change such as in industry, leisure and agriculture	Knowledge and understanding of fragile landscapes such as deserts, polar regions, mountains and reefs	Asking questions about place
Knowledge of Ireland, Britain/UK, in the context of Brexit, Europe and the European Union	Knowledge and understanding of resource distribution and food, water and energy security on regional, national and international scales	Understanding different approaches to managing and living with changing physical and human environments	Assessing positive and negative aspects of place along with current and future opportunities and threats
Broad knowledge of the world including continents, oceans, countries, significant features of Earth such as wind patterns and tectonic structures	Understanding the reasons for and processes behind the location and changing distributions of population	Knowledge and understanding of processes involved in distribution and patterns of major physical features, including natural regions and ecosystems	Exploring how place is portrayed in media including local and international newspapers, TV programmes, films and social media
Knowledge and understanding of specific places or regions different from their own, focusing on people–environment interactions	Understanding of flows and movements of people, goods and ideas, with examples on a regional, national and global scale	Knowledge and understanding of the Earth's oceans and their significance	Assessing other people's opinions about place
			Maps – what they show us, how to use them and how to construct them
			How to use other sources – photographs, diagrams, internet,

(Continued)

12 Powerful primary geography

Table 1.1 (Continued)

Place (places, territories and regions)	Space (patterns and links)	Environment (physical and human interaction)	Geographical enquiry (procedures and tools)
Knowledge and understanding of places of great significance in and for the world today Knowledge and understanding of places that are scenes of conflict at different scales (e.g. a local place, Afghanistan) Knowledge and understanding of places where physical extremes or hazards dominate Understanding that people have different perspectives and perceptions of places	Understanding of spatial systems, such as climate, through the distribution of energy by ocean currents and wind patterns Knowledge and understanding of issues that arise from uneven distributions of people and wealth Understanding the role of imagination and speculation in envisioning alternative uses of space in the future	Understanding landscapes as distinctive collections of landforms, soils and earth surface processes Understanding the links between social, economic and environmental quality Understanding renewable and non-renewable resources from the Earth and its atmosphere Understanding systems-thinking in the context of human and physical environments	databases, animation and visualisation technologies, electronic atlases, film libraries, newspapers, magazines and journals, etc. First-hand investigation via fieldwork, photography, GPS sketching, interviewing, meeting people, etc. Writing descriptively and analytically about places, spaces and environments; constructing and challenging arguments

Source: Geographical Association, 2011: 11

Case study 1.1 Exploring the geographies of everyday items

The big ideas of geography can be explored through the products children have in their possession. Tracing a product's geographical connections teaches children how products are made and transported (Figure 1.1). It highlights the geographical and procedural journeys from component materials to final products. Children from Scoil Íde, Corbally Limerick (9–10 years), traced the global connections of familiar everyday objects such as a hockey stick, a hoodie and a pencil. Through research they compiled a commodity journey log for their chosen item. Each child made a presentation to a wider audience in school and received feedback. Resources such as the Sourcemap website allow children to visualise the journey of several commodities in graphic formats. Applying a 'follow the thing' approach (Cook, 2004; Cook et al., 2017), children used questions (Table 1.2) to guide their research:

Table 1.2 Research questions for researching the geographies of everyday items

Choose any object/food/material of interest to you and research it.
Title: The geography of (my chosen item)

What are the raw materials, i.e. what is it made from?

Where can they be sourced? Show this on a map.

How is your item transported, e.g. boat, train, truck, etc.?

Is it imported/exported to/from Ireland?

How much does it cost to make/produce the item?

What is the impact on the environment of the production of your item, i.e. pollution, and use of fossil fuels, natural resources, water, etc.?

Can it be recycled?

Where is it produced?

What are the workers paid? Is it a fair wage?

Requirement: Use PowerPoint, Prezi or any other presentation software including Microsoft Word.

Include pictures, maps/statistics/facts and any other material which may be relevant.

Acknowledge your sources.

Figure 1.1 The geographies of everyday objects

Once children have researched and presented the geography of their chosen item, children can reflect on the visible and invisible impacts of their items on local and global communities. Videos such as *The Story of Stuff* (https://storyofstuff.org) produced by Free Range Studios with Annie Leonard offer an accessible examination of our relationship with materials and consumerism. From its extraction through sale, use and disposal, all the stuff we buy and use affects communities at home and abroad, yet most of this impact is hidden from view. Based on Leonard's research (2010), *The Story of Stuff* exposes the connections between environmental and social issues, and calls on all of us to create a more sustainable and just world. It helps us to think about our purchase, use and disposal of stuff.

Geography: a powerful discipline

Geography is a powerful discipline. It has the power to inform and transform, to educate and to challenge prior ideas; to inspire and to motivate children to be creative. Ultimately, it equips children with conceptual understanding and knowledge to enable them to live in, to comprehend and to participate in their world. This discipline helps us to understand change, conflict and important issues which have an impact on our lives today and which will affect our future lives. The disciplinary lens of geography provides teachers with a powerful resource for learning.

According to UNESCO (2012) educational initiatives need to address current environmental, social and political challenges facing our world. Such initiatives require holistic and interdisciplinary approaches. The four pillars of education for all – learning to know; learning to do; learning to live together and with others; and learning to be (International Commission on Education for the 21st Century and Delors, 1996) – provide a useful model for thinking about quality education in general and quality geographical education in particular. Powerful geography is a curricular area in its own right. But it also has to be cross-curricular to be truly effective. It does not just refer to knowledge, powerful or otherwise; it is a holistic approach which applies to skill development and affective responses. Effective primary teachers can help children develop geographical conceptual knowledge throughout the day and across the curriculum. While many teachers are struggling to cope with an overcrowded curriculum, geographical understanding can be developed through literacy, numeracy and other curricular areas in a meaningful way for children.

Powerful geography does not represent a box of geographical facts handed over to children, but rather a process of collaborative learning whereby concepts are explored, examined, showcased and debated. Children learn best when they are fully engaged in their learning, when what they learn matters to them and when they have a sense of ownership and agency over the process. Catling and Martin suggest that 'children's understanding and knowledge is powerful and to be valued, that it has rationality, is systematic and structured, has coherence and is conceptually grounded' (2011: 9).

If geography is so powerful where is the universal demand for the discipline in our schools? Unfortunately, the public image of geography remains somewhat constrained. In some schools the potential of geography is unrecognised. In England, the humanities, including geography, are struggling to obtain the curriculum time deserved (Barnes and Scoffham, 2017). In the USA, geography is taught as part of the social studies curriculum, the marginalisation of which has been well documented (Heafner and Fitchett, 2012). Internationally, there has been an increased focus on literacy and numeracy. Ironically, denying children the opportunity to build geographical vocabulary and background knowledge actually leads to lower literacy levels (Grant, 2007; Statment, 2009). In a globalised society, which requires collaborative problem-solving skills and critical thinkers to address complex social economic and environmental concerns, core geographical education is as important as literacy, numeracy and computational skills.

The relevance of geography is ubiquitous. One only has to open a newspaper or follow a news report to appreciate the importance of geography. As children progress through the school year, seasonal events, national celebrations and thematic events offer numerous opportunities for geographical explorations.

Of course there is the issue of quality in primary geography and what exactly constitutes excellent primary geography teaching. Children have mixed experiences ranging from inspirational thought-provoking geography to low-key minimalist approaches. Catling (2017) examined reports from the four jurisdictions of the UK to explore the issue of quality in the teaching of humanities in primary schools. According to his findings, primary teachers need to be knowledgeable about the discipline-specific content of geography, history and religious education and able to see interlinkages between the three areas. High-quality teaching requires a substantial degree of pedagogic content knowledge (PCK) (Shulman, 2003). This includes curriculum knowledge, pedagogical knowledge, understanding of learners and knowledge of resources. According to Catling (2017: 363) high-quality teaching

> emphasises holding high expectations of children's learning, drawing on a wide repertoire of teaching strategies and approaches, using active and enquiry methods, appreciating the nature and range of resources that can be used, and understanding how best to enable learning for each child.

Other features include teachers' enthusiasm, flexibility and the willingness to take a risk. While these features underpin all high-quality teaching (James and Pollard, 2012), Catling makes the point that context matters and that the importance of humanities including geography needs to be valued and appreciated by children, teachers, schools and inspectors.

Current context

In these uncertain times, geography matters more in the 21st century than heretofore. To date, the 21st century has been characterised by rapid and unpredictable change. World development indicators prepared by the World Bank (wdi.worldbank.org) demonstrate an unequal distribution of the Earth's resources and increased pressure on the world's fragile ecosystem. This century, there has been a marked increase in fundamentalism and belief-based conflicts, with the threat of terrorism having a marked impact on security, tourism and travel. Numerous natural disasters including drought, flooding, earthquakes and volcanoes have also occurred. Craft (2012: 173) refers to this as a time of 'immense human over-confidence alongside increasing instability in the economy, social structures, beliefs and environment'. In other words we are witnessing an increased disconnect between humanity and the Earth upon which its existence is based.

Our global safety net, biodiversity (otherwise known as the variety of living things on Earth), is facing unprecedented challenges. As humans, we depend on the Earth's biodiversity for food, drinking water, clean air, medicines and shelter. According to the Living Planet Index (WWF, 2018) unsustainable human activity is pushing the planet's environmental resources beyond its capacity. Wildlife populations around the world have fallen by an average of 60% over the last 40 years, according to the WWF study of thousands of vertebrate species. As Hicks (2014) asserts, we are educating children in 'troubled times'.

Extensive flooding in the UK and Ireland together with the destructive power of hurricanes in the Caribbean have highlighted changing climatic patterns and global warming. Devasting forest fires which occurred in Australia, California and the Amazon rainforest during 2019, together with accelerated Arctic and Antarctic warming, underline the international challenges posed by climate change. The devastation caused by several earthquakes, their coverage in international media and the world's reaction have further demonstrated the power of geography. Equally, war and conflict in countries and regions such as Syria, South Sudan and the Middle East cast a light on unacceptable levels of human suffering. Water shortages, famine, migration of people, disputes over natural resources, complexities associated with world trade, interdependence, globalisation and debt are all international challenges with which our world is grappling today.

An unequal distribution of wealth both within and between nations, exacerbated during periods of recession, raises important geographical questions. Living standards, levels of equality and inequality, patterns of production, movement of people and goods are still largely determined by geographical factors. Place of birth matters. For instance, those born in Western Europe and the USA enjoy the highest standards of living in the world. Inequality is extreme, rising, and widely accepted to be a threat to us all and to the global economy. Manifesting itself in many different forms, inequality is a political, social and economic reality in society today.

We are living in a time of unprecedented change. Some refer to this as the Fourth Industrial Revolution (Schwab, 2017), an era driven by new technologies, globalisation and automation. Digital technology is having a profound impact on society, education, schools and children. Google has changed our world and our access to geography. Instant access to the internet is available through smartphones, iPads, laptop computers and gaming devices. Virtual globes such as Google Earth allow children and teachers to view their area at different scales from vertical, oblique or three-dimensional (3-D) perspectives. Facilities such as Street Maps provide 360° panoramic views. Google Earth allows users to input their own data within certain parameters. Digital technology is allowing us to make connections across time zones and places in a way that was previously unimagined (Figure 1.2).

The process by which the world is becoming increasingly interconnected through increased trade and free movement of people, culture and ideas is commonly referred to as globalisation. Now facing a backlash, the future of economic globalisation is somewhat uncertain. Significant political events including Brexit, the growth of right-wing movements in Europe, and the election of President Trump collectively highlight increased support for restricting movement of people. The fall of the Berlin Wall in 1989 heralded a time of great hope. Today we seem more divided than ever as is apparent in a global preoccupation with walls and borders. According to Marshall (2018: 2) 'at least 65 countries, more than a third of the world's nation states, have built barriers along their borders; half of those erected since the Second World War sprang up between 2000 and now.' The people most negatively affected by these walls are the poorest and most marginalised in society. Traditional global power dynamics are changing. We are living in an 'era of democratic recession' when the world is becoming more authoritarian than democratic (Doucet and Evers 2018: 3).

Today, there are refugees in countries all over the world. Many people are not able to avail themselves of the protection of their state and therefore require the protection of the global community. Over 65 million people worldwide have been forced to flee violence,

Figure 1.2 Exploring geography through digital technology

conflict and persecution while millions more have left everything behind following natural disasters. According to the UN Refugee Agency there are more than 15 million refugees in the world today. Refugees are a reminder of the failure of societies to exist in peace and our responsibility to help those forced to flee. Indeed, the origins of much of the world's political strife can be traced back to the colonial expansion and empire building of European nations. Now asylum seekers are seeking refuge from wars initially caused by European or American interventions whether through bankrolling dictators, supplying weapons, participating in wars for financial gain or creating contentious borders. The West, which has disproportionately benefited from globalisation, refuses to accept its responsibilities towards asylum seekers, even though the current mass movement of people is a direct result of the greed inherent in a global capitalist system.

Nevertheless there are reasons for hope and optimism. The world is witnessing substantial progress, achieved in part by the Millennium Development Goals. Building on these goals, the United Nations has adopted the Agenda for Sustainable Development, which includes a set of 17 Sustainable Development Goals (SDGs) to end poverty, fight inequality and injustice, and tackle climate change by 2030. There are many resources available to help children understand the concept of inequality, the unequal distribution of resources, and the impact of inequality. Envisioning alternative ways of sharing resources and living sustainably provide a hopeful focus for children. The significance of the SDGs is discussed in greater detail in Chapter 8.

It is important for children and young people to learn about the world upon which they depend. Critical skills of analysis are essential requirements for assessing multiple perspectives. It is equally important for young citizens to learn to appreciate diversity, become aware of environmental issues, and learn how to promote sustainable lifestyles. As the technological landscape changes so too does childhood. Children now have increased access to the internet; many have a parallel existence in virtual space. Craft (2011: 174) describes children as 'skillful collaborators, capable of knowledge-making as well as information-seeking'. Children are engaging in social networks, they are learning to create and navigate digital content. While this comes with freedoms and dangers, potential and risks, there are significant opportunities for the teaching of powerful primary geography. For the first time in human history we understand the impact of our actions on the environment. We are already exploring new ways to feed our growing population, meet our energy demands and manage our water supply. Notwithstanding the deniers, we have the knowledge and capacity to move towards a better future for people, biodiversity and the climate. Powerful primary geography includes children's voice as part of this agenda.

Developing 21st-century learning, competencies and skills

Today, globalisation, technology and the growth of knowledge are three of the major forces of change shaping the world our children inhabit. We live in a globalised and interconnected world facing numerous environmental, economic, social and political challenges. The 21st-century learner will sell to the world, buy from the world, work for global companies, work with people from different cultures, collaborate with people all over the world and solve global problems. For many years, educators have been engaged in a reassessment of the knowledge, skills and dispositions young people need for success in today's rapidly changing and complex world. In 2018, the Organisation for Economic Co-operation and Development (OECD) launched a new assessment of global competence as part of PISA (Programme for International Student Assessment). This followed the launch of the SDGs, which officially recognise the importance of education for global citizenship. Both of these noteworthy developments are explored in greater detail in Chapter 8.

Living in a competitive, globally connected and technologically intensive world requires problem-solvers and critical thinkers. While problem-solving, critical and creative thinking have always been part of the educational process, these competencies are now considered to be core in 21st-century learning (Trilling and Fadel, 2009: 50). Twenty-first-century competencies include critical thinking, problem-solving, communication, collaboration (teamwork), creativity and innovation, knowledge building (a growth mindset), resilience, citizenship and global competency (Trilling and Fadel, 2009: 50; Schleicher, 2018; Colvin and Edwards, 2018). These competencies are considered essential to help children and young people solve complex, messy problems including those that have yet to be encountered. Competencies are broader and more complex than knowledge and skills. According to the OECD a competency

> involves the ability to meet complex demands, by drawing on and mobilising psychosocial resources (including skills and attitudes) in a particular context. For example, the ability to communicate effectively is a competence that may draw on an individual's knowledge of language, practical IT skills and attitudes towards those with whom he or she is communicating.
>
> (OECD, 2003: 4)

Just as technologies are rapidly expanding, so too is the volume of knowledge. In 24 hours the average person generates as much data as a person did during his or her lifetime 10 years ago (Kennedy and Murphy, 2017). Hence, children need to be proficient in data handling skills. These include: asking questions, deciding what data they need to answer these questions, locating the best sources for this data, filtering different types of data and presenting data in a meaningful and engaging way to a wider audience. Children need to learn the skills of scrutinising data carefully. Critical thinking encourages children to ask questions, explore the hidden meaning of data and be open to alternative perspectives. This book argues that primary geography is well placed to help children develop these core 21st-century competencies with a view to creating a more sustainable, more caring and more democratic society (Biesta, 2015).

The political power of geography

The political power of geography is well documented (Dorling, 2017). Geography allows us to compare places according to different criteria, to see what is possible elsewhere and potentially to make changes at home. Politicians travel to other jurisdictions to see how issues of common concern are faced. Nonetheless, those whose interests are served by inequality or who are in favour of maintaining the status quo have long realised the danger of teaching geography. In testimony to a Select Committee of the House of Commons in 1879 concerning the expenditure of the London Schools Board, one petitioner declared: 'Geography, sir, is ruinous in its effects on the lower classes. Reading, writing and arithmetic are comparatively safe, but geography invariably leads to revolution' (Independent, 2002).

Deciding what geographical content and skills to teach and how to teach are fundamentally political decisions. Teachers can choose to teach about people and places in a safe manner. Alternatively teachers can encourage children to ask questions, dig deeper and critically interrogate some of the dominant geographical messages portrayed in society today. Owens (2013: 392) refers to the 'messy' nature of geography as it sometimes deals with contentious issues, local and global topics, that may be considered sensitive or controversial. Controversial issues include questions or problems which generate different opinions. They can include issues with political, economic, social or environmental impacts on children, their families and the wider community (locally, nationally or internationally). Such issues have many viewpoints ultimately informed by personal and collective values and/or beliefs.

Mass media and increased access to social media have exposed children to sensitive topics that require informed discussion in the classroom (Scarratt and Davison, 2012).

Geographical investigations and 21st-century competencies such as critical thinking, collaboration and creativity equip children to navigate the multiple messages which invade their lives today.

Powerful geographical knowledge

Geography, both as a discipline and a body of knowledge, has been debated by several commentators. Traditionally, geographical knowledge was conceptualised as a fixed body of objective knowledge to be passed from teacher to child. However, research on the sociology of knowledge demonstrates that geographical knowledge is socially constructed. Instead of one, there are several views of the world. Geographical knowledge is fragmented and incomplete.

Young (2008) coined the term 'powerful knowledge' as part of his argument for a subject-based curriculum. He has argued that while knowledge is a social product, the knowledge found in academic disciplines such as geography has real features, as it represents the stored accumulated knowledge and understanding of a community of researchers. It is this powerful knowledge that allows children to transcend and go beyond their everyday experiences. For Young, powerful knowledge 'means knowledge that is reliable, fallible and potentially testable' (2008: 182); this suggests a social realist view of knowledge which contends the existence of an objective reality. Young's ideas have been developed further through the Geo-capabilities Project, where geographical knowledge is understood as 'Powerful Disciplinary Knowledge' (PDK) (Lambert et al., 2015). The geo-capabilities project suggests acquiring 'powerful' geographical knowledge through geographical education is essential for developing 'capability'. Also influenced by Nussbaum and Sen (1993), the geo-capabilities approach emphasises the value and role of geography as a school subject in cultivating the development of human capabilities.

This articulation of powerful knowledge has been critiqued by those holding a constructionist or relativist view of knowledge (Roberts, 2017). Constructivist pedagogy is rooted in the idea that knowledge is not an objective construct but emerges as we engage with and experience reality. Each of us perceives and understands the world differently depending on prior knowledge, former experiences and thinking frameworks. Constructivism emphasises the importance of the teaching context, children's prior knowledge, active engagement of the child and interaction between the child and the emerging knowledge.

I concur with Huckle (2019) when he argues that geographical knowledge is powerful if it is critical and empowering. However, this idea of powerful knowledge needs to be framed in terms of Catling's empowering pedagogy (2014). Catling's work on empowering pedagogy underlines the importance of children's voice and agency. As the world becomes increasingly complex and due to the contested nature of geographical knowledge, empowering pedagogy is what (according to Catling, 2014) gives geography its power. Children can be co-teachers (Catling, 2014) in a classroom where knowledge resides with all, not with the teacher alone. Craft (2011) argues that teacher-centred approaches and a curriculum emphasising content is at odds with the educational needs of children and the nature of childhood. An overemphasis on knowledge for its own sake could provoke a return to a curriculum of facts and figures.

What might powerful primary geography look like?

What is knowledge and what does it mean to know? These are philosophical questions explored by educators for centuries. The nature of knowledge and knowing are constantly changing. Knowledge is widely available and cheap to access. YouTube provides instant answers to daily queries. Millions of people are accessing Massive Open Online Courses (MOOCs) offered by universities around the world. Nonetheless, the importance of understanding how knowledge is generated, how discipline-specific knowledge connects with other disciplines, and how to apply knowledge to solve problems is more important than heretofore. Twenty-first-century competencies of collaboration, communication, creativity and problem-solving are recognised as essential in a preparation for life, not just exams. While geography is considered by some 'the essential skill for the 21st century' (Nagel, 2008), it has the potential to play a vital role in helping children develop 21st-century competencies.

Geographical knowledge is contested. Debates in the literature about geographical education centre on the nature of powerful knowledge, its meaning and role in school geography (Young et al., 2014; Maude, 2016, 2018; Catling, 2018). Because of the challenges facing society today, I suggest that any framework of powerful knowledge must be considered along with 21st-century competencies. Table 1.3 charts a picture of powerful primary geography. It lists five types of powerful geographical knowledge (Maude, 2016), alongside 21st-century competencies, together with attributes of powerful primary geography further explored in the indicated chapters of this book. While Maude was not

Table 1.3 Charting a picture of powerful primary geography

Maude's five types of powerful knowledge	21st-century competencies	Teaching powerful primary geography
Knowledge that provides children with new ways of looking at the world	Knowledge building	Thinking geographically (using key concepts such as place, space and the environment), i.e. testing and evaluating claims about knowledge, enabling thinking in the subject Chapters 2 and 3
Knowledge that provides children with powerful ways to analyse, explain and understand the world	Critical thinking	Asking geographical questions (making connections, understanding interconnections and developing well-informed geographical understanding) Chapters 2 and 3

Knowledge that gives children some power over their own knowledge	Real-world problem-solving	Formulating geographical solutions, i.e. using and applying the subject to contribute to topical and societal matters Chapters 3, 4 and 7
Knowledge that enables young people to follow and participate in debates on significant local, national and global issues	Collaboration and teamwork	Acting geographically, extending information and understanding about the world's environments, places and people Chapters 5 and 6
Knowledge of the world	Global competency	Becoming an ambassador for geography Chapters 7 and 8

necessarily referring to primary geography, his categories of geographical knowledge have been adapted for a primary classroom.

The aesthetic aspect of powerful geography

The word 'aesthetic' comes from the Greek word *aisthetikos*, meaning the perceiving of things through the senses, which includes sensory and perceptual dimensions. An expression of inner feelings and experiences is the sensory dimension while the perceptual dimension refers to the creation of a response, e.g. a drawing based on personal reflections and interpretations. Perceiving through the senses is influenced by the physical manipulation of certain materials, whether this occurs through writing, drawing or playing music. The Swiss educationalist Johann Pestalozzi described aesthetic production as the link between head, heart and hand (Heafford, 2016).

To engage children in geographical learning, the affective domain, which includes emotions, attitudes and motivation, should be engaged. Geography has a particular contribution to make to emotional development and emotional literacy because it involves the study of real people, places and issues (Tanner, 2010). Emotionally literate geography education enables children to express their own feelings; recognise how they are connected to a place; recognise how they feel about a place; understand the feelings of others towards a place; and communicate responses in different ways. According to Tanner (2010: 37) geography offers three major opportunities for the development of emotional literacy:

- It helps children to recognise and express emotions associated with places and environmental issues;
- It provides opportunities to develop empathetic understanding of others' feelings and views; and
- It develops interpersonal skills through the active learning approaches required by meaningful geographical enquiry.

Geographical experiences have the potential to contribute to children's wellbeing. Children's attachment to place can be fostered though creative place making and place-exploration activities such as field trips, orienteering and den building. These activities are holistic in that they nurture personal, social and emotional development.

Awe and wonder

One of the things I enjoyed most as a primary teacher was experiencing the awe and wonder of children as they witnessed the miracles of nature occurring in their local places (Figure 1.3). Children came to me with expressions of joy as they showed me the biggest acorn ever or described the most beautiful moon witnessed the night before.

Awe is a universal experience and part of our innate DNA. Many famous travel writers including Elizabeth Gilbert, Bill Bryson, Paul Theroux and Ernest Hemingway convey a sense of awe and wonder in their writings. As children the world is an awe-some and wonder-full place. Piff et al. (2015) describe awe as 'that sense of wonder we feel in the presence of something vast that transcends our understanding of the world'.

Being in nature and experiencing an ongoing sense of awe and wonder is important for our sense of happiness and wellbeing. Awe and wonder can be described as an emotional response, a sense of 'wow' which one experiences when exposed to beauty in nature, panoramic views, spiritual experiences, existential experiences, inspiring music and art. Feelings of elation and freedom can be triggered as a result. Piff et al. (2015) found that experiencing a sense of awe promotes altruism, loving-kindness and magnanimous behaviour.

Scoffham (2016) addresses the idea of our oneness with nature. If we saw ourselves as connected and part of a bigger reality would we live differently? Would we spend more time outside soaking in the benefits of nature? Drawing on ideas from Froebel, Jung, Otto, Capra and Luisi, Scoffham emphasises how our sense of oneness with nature can awaken a sense of awe and wonder. He argues that telling the 'story of the world' needs to draw on multiple perspectives and that emotional encounters and existential moments are a necessary part of a meaningful geography curriculum.

Teaching geography powerfully nurtures children's natural sense of awe and wonder. In a study of 134 children from schools in 5 different areas of England, a significant proportion of questions the children asked about geography were underpinned by a sense of awe and wonder (Scoffham, 2013).

Figure 1.3 Children exploring their environments to discover hidden treasures

> *Exercise 1.1 Generating a sense of awe and wonder*
>
> 1 Using your camera or smartphone, take one photograph of something which inspires awe and wonder for you. Try to share an image which made you say 'wow'.
> 2 Share these photos with your staff and discuss the special elements and unique features of each photograph.
> 3 Discuss ideas for promoting and sharing children's awe and wonder.

Empathy and responsibility

Once children become involved in making choices and decisions, the question of values arises. By making decisions and articulating reasons for their choices, children can begin to make sense of the world around them. Many solutions are based on a compromise and ultimately go against some people's wishes while fulfilling those of others. Children need to understand that issues are complex and decision-makers do not automatically reach the correct or incorrect decision. Powerful geography which involves children in making decisions generates a respect for evidence, an awareness of biased information and an interest in contemporary issues. By teaching how to appreciate alternative viewpoints, powerful geography can develop empathy in children. It is important to understand that simple explanations rarely tell the complete story.

The issue of child agency is important in geography. Geography informed by citizenship education helps children to understand and participate in the environment in which they live. Such participation implies taking responsibility and showing care. Such education initiatives give children a voice and create a climate in school where children can talk about issues. Helping children take responsibility for their learning is a good place to begin.

Powerful geographical thinking, enquiry and skill development

Biesta (2015: 2) explores the concept of education as an encounter between the child and the world through which the child comes into the world and acquires a 'worldly' form so to speak. In this conceptualisation of education, both the child and the world are changed as a result of the encounter. From a geographical perspective, the idea of an encounter with the world is a wonderful analogy for planning enquiries and investigations. Through these investigations and enquiries teachers need to explicitly teach children how to think geographically. By using discipline-specific concepts and various thinking frameworks, children become more confident in the use of geographical language, in their conceptual understanding and general thinking ability. The benefits of enquiry-based learning and thinking skills are explored in greater detail in Chapter 2.

Powerful conceptual knowledge is formulated through investigative and enquiry-based learning. Based on teachings of Dewey, Bruner, Gardner and Freire, geographical enquiry involves the development of investigation skills inside and outside the classroom. Generally, based on a series of questions or one big question, children undertake a series of investigations, develop subject-specific skills as well as cross-curricular skills, present their findings using a variety of media and reflect on the learning experience.

Skill development is a key component of geographical education. Geographers work in a variety of important occupations including researchers, teachers, journalists, cartographers, tourism advisers, environmental managers, hydrologists, urban planners, community development officers and seismologists. All geographers ask questions, investigate, explore and conduct enquiries.

The idea of children 'doing' geography as opposed to 'learning' geography is a useful metaphor for teachers. Thinking about child engagement and what the child will be doing at different stages of the lesson or geographical enquiry moves the focus from the teacher to the child. As Gardner (1999: 108) suggests:

> The brain learns best and retains most when the organism is actively involved in exploring physical sites and materials and asking questions to which it actually craves the answer. Merely passive experiences tend to attenuate and have little lasting impact.

Case study 1.2 The Keep on Track Project

Sometimes the best ideas are the simplest. Take the railway line from Dublin to Galway or indeed any railway line. What if primary schools adopted each railway station along the line and conducted a cross-curricular investigation into each local geographical site? What if each school could publish its work and share progress with other schools involved in this work? This is exactly what happened during the *Keep on Track Project*. This collaborative cross-curricular project connected 16 primary schools from Galway to Dublin along the Inter-City rail line. The children in each school adopted their local station and researched its story. Classes from senior infants to sixth class (5–13years) took part in the project, researching materials and engaging in learning experiences relevant to their interests and age group. This project was the brainchild of primary teachers Kate Murray and Cathal O'Conaill, and was supported by teachers in each of the participating schools. The project was a great model of continuing professional development (CPD) for teachers as it facilitated the sharing of practical ideas, concepts and curriculum implementation among peers in a supportive manner.

Children from the schools made connections with each other and shared their work through use of Web 2.0 tools such as Skype, Twitter and a group blog. This project engaged children with their curricular subjects in a motivating manner and offered opportunities to connect with and learn from other schools. Keep on Track offered opportunities for teachers and children to engage with 'real-life' learning experiences in a cross-curricular manner. It also strengthened collaboration between schools, Irish Rail, Education Centres and the National Centre for Technology in Education, and provided a model for cooperation between schools and industry. Schools compiled their work across all curricular areas and presented project materials and outcomes on a joint project blog, www.keepontrack.scoilnet.ie.

Irish Rail (www.irishrail.ie) made archive materials available to aid the implementation of the project and facilitated visits to local railway stations. Children documented their visit and collected data through photographs, interviews and on-site observations. Staff from each station visited primary schools to answer children's questions.

The Galway Schools Library Service offered its services to the project and provided sets of class novels such as *The Railway Children* by Edith Nesbit and *Stop the Train* by Geraldine McCaughrean and research materials to the schools involved. This allowed schools to operate online book clubs to discuss and explore the novels.

A launch day was held in a hotel in Galway where schools presented their project findings. A special Irish Rail train brought all schools involved to and from Galway for the day of celebration. Each class set up a stall and made a presentation based on their research. The stalls were set out in line with the stations from Galway to Dublin so walking through the hall was akin to virtually travelling the journey by train (Figure 1.4). This project could be replicated linking schools on a railway line, a river or locations in a particular geographical area. All participants were extremely positive as illustrated in the feedback below.

This project was chosen to represent Ireland at the Microsoft Innovative Education Forum in Moscow. The Forum, which has been in operation since 2004, is an international gathering of exceptional educators presenting their learning projects achieved through innovative usage of technology. The purpose of this event is to promote international sharing of innovative practices of technology integration and provide Europe-wide networking opportunities among top innovators. It also focuses international attention on the importance of technology innovation in education.

Feedback from project participants

> I have been involved in a lot of projects where schools work co-operatively and I am a big fan of web 2.0 technology such as blogging, podcasting and film making and getting children working hands on and using these technologies as a method. So I thought it would be a great idea to link the schools from Dublin to Galway. Literacy and numeracy were integral to the full project. First of all it's a blog so all the children's work was recorded in writing on the blog or else they used podcasting methods. Their oral language was developed, their written work was developed. They also engaged with Galway Library services here in Galway and the libraries across the country.
>
> Kate Murray, Principal, St Augustine's NS, Clontuskert, Co. Galway

> It's a very innovative idea. Technology links communities near and far across the globe indeed but in this particular project across the country. The railways provide a physical link for the same communities. From an Irish Rail perspective it's great to be involved with the young people some of whom never travelled on a train before this project. We see children as the rail users of the future.
>
> Gerry Glynn, district manager Galway, Irish Rail

> Keep on Track was a great example of how the teaching of geography allows creative opportunities for a cross-curricular approach, cooperation between schools and their localities, and above all generates excitement and enthusiasm for young people. In this project Geography, as a subject, allowed every young person the opportunity to display their particular skills and talents.
>
> Bernard Kirk, director Galway Education Centre

Reactions from children

> I learnt so much about railway travel from this project.
>
> Travelling on the train was the best bit!
>
> I loved the show in the Radisson where we saw the work from other schools.
>
> My cousin was working on the Tullamore project so we had a chance to share our ideas.

26 *Powerful primary geography*

Figure 1.4 Picture from Keep on Track (more images are available in Section 7 of the colour plates)

Case study 1.3 The geography of a bridge

Bridges are everywhere; over roads, rivers and canals, spanning estuaries and joining islands to the mainland. As significant geographical features of urban and rural landscapes, bridges are very important for connecting places and allowing us to travel easily. These incredible architectural feats impress and delight us. An example of human-made features, they illustrate the impact humans have on the local environment.

The word 'bridge' is commonly found in Irish place names. When located at the end of a name, this means that the town developed beside a bridge, or became famous as a bridging point. The name of the village of *Clarinbridge* comes from the original plank bridge across the river Clarin. Other place names are linked with industry and engineering. Ironbridge is a settlement on the river Severn in Shropshire. The village developed beside, and takes its name from, the cast-iron bridge built there during the Industrial Revolution. Telford in Shropshire was a planned New Town in 1963 to draw off population and industry from Birmingham, and is named after Thomas Telford, the famous civil engineer. Indeed, the town of Bridgnorth is named after one of the bridges he built.

Children (8–9 years) from Gael Scoil de hÍde, Oranmore, Co. Galway explored bridges in their local area along the river Clarin and the river Clare. They visited their local bridges, took photographs, made sketches and mapped the location of the bridges. Children recorded their own thoughts about each bridge and used these notes later as material for poetry and art. They discussed the impact of each bridge on the local area especially in the context of extensive flooding which had occurred recently. Children played pooh sticks on the bridge and observed the flow of the water in the river.

Civil engineers plan and design bridges. Working as engineers, children from Gael Scoil de hÍde completed a geography and engineering project about bridges. The children examined a number of local and international bridges exploring the following enquiry questions:

> What is a bridge? Where would you find a bridge? Is there a bridge in our locality? Where can you find significant bridges in Ireland? (other countries?) What are bridges made from? What do they all have in common? What are some differences? Which

bridge is the strongest? What makes a good bridge? Look at the shapes that are used to construct each bridge. Which shape is the strongest? Which is the weakest? Why might each type of shape be used to build a bridge? What types of bridges are there? Can you name any? (Bridge designs include arch, suspension, etc.)

Drawing inspiration from local bridges, children worked in groups to design their own models, recording bridge type and elements, maximum weight, weaknesses and strengths. This bridge planning, designing and engineering activity provided opportunities for children to be creative, to explore different materials, test ideas and problem-solve. When children designed their bridge, they had to consider how it would be used, how long and wide it should be, and how much weight it could hold. The following factors were also taken into consideration by the children:

- Who will be using the bridge? Cars? Pedestrians? Bicyclists? Big trucks? All vehicles?
- What materials should be used to build the bridge?
- The triangle shape is popular in the construction of bridges. Can you find examples of the use of this shape in famous bridges?
- What style should the bridge be?
- In the case of a river, how long must the bridge be? What is the water like?
- What is the land around the water like? Rocky, muddy, sandy?
- What bridges are used locally?
- What inspiration can be found in bridges around the world?

Children studied examples of the following bridges:

Beam bridges made of horizontal beams supported by piers at each end.
Truss bridges using a combination of triangles made of steel.
Arch bridges using arches to support the bridge.
Suspension bridges such as the Golden Gate Bridge.

The children presented their projects at the Galway Science Fair as illustrated in Figure 1.5.

Figure 1.5 Pictures from the geography of bridges

The power of personal geographies

According to Catling and Martin 'children construct their own ethno-geography' (2011: 8) of places, spaces and environments from life experience. Ethnogeography (Martin, 2005, 2008) reflects the view that all learners (including children) are geographers because of their lived experiences. Drawing on the concept of children's geographies or children's geographical knowledge can change the way geography is taught in the 21st century (Catling, 2011). It ensures that children are active agents in their own geographical learning. It offers teachers a way to engage with their own personal geographical learning as well as that of the children in their class. This assists teachers to enhance the curriculum and to become 'curriculum makers' through reconstructing, revitalising and owning the geography programmes in their schools. According to Catling (2011: 27) this enables teachers to develop their own geographical studies to foster, excite and deepen children's geographical understanding and engagement in and beyond school.

Teachers also come to the classroom with their own lived geographies. Geography is part of everybody's lives because 'we all live in the world' (Martin, 2006: 1). This is supported by Morgan (2006) who suggests that the experience of living in the world makes geographers of us all. Geographical knowledge is constructed by individual learners, depending on perspective, previous geographical experiences and learning styles. Teachers' everyday geographical engagement comprises places (Where do I come from? Where do I live? What is my connection with the place where I teach?), daily travel and interconnections with their human and natural environment. Teachers also bring their own memories of geographical engagement from previous formal and non-formal educational and life experiences (Figure 1.6). Just as children's geographies are often unrecognised as a reservoir of resources for geographical engagement so too are teachers' previous geographical experiences. A teacher's personal geographical experience shared with children is ultimately much richer and more meaningful to children than any chapter from a textbook. Martin's concept of ethnogeography illustrates how teachers' engagement with geographical ideas and experiences in their daily lives is an important part of their pedagogical and geographical frame of reference.

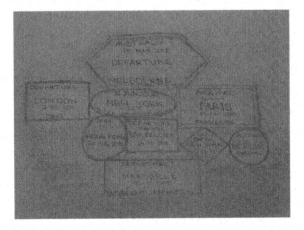

Figure 1.6 Example of one teacher's lived geography

> *Exercise 1.2 Personal reflection for teachers*
>
> Where are your parents from?
>
> Where were you born?
>
> Where are your family members currently living?
>
> Where have you travelled within and outside your country?
>
> Where is your favourite place locally, nationally and globally? Reasons why?
>
> Where would you like to go and why?
>
> How can your personal geographical knowledge become a resource for you as a teacher?

Conclusion

Powerful primary geography has the potential to inspire children, student teachers and teachers to become passionate about their world and their place in this world. Through my research for this book I have been humbled and encouraged by the wonderful geography I have witnessed in primary schools. These examples of primary geography are meaningful, allowing children to make real connections, thus expanding their conceptual understanding and skill development. As geographers, children can engage with their world in a myriad of interesting, educational and challenging ways while enjoying themselves at the same time.

Children are living in an exciting world. They have extraordinary questions. It is our job as teachers and educators to place children and their questions at the centre of geographical exploration in the classroom. To ignite their passion and curiosity, as teachers we need to focus on ourselves first, renewing in ourselves an awareness of the miraculous and magical nature of the world around us.

The ability to make decisions independently and collaboratively is an important 21st-century skill. The act of engaging in geography requires us to think creatively. Roberts (2017) argues that geographical education is only powerful if it adopts a powerful pedagogy that values children's everyday experiences, promotes critical thinking and actively involves children through their construction of knowledge and understanding.

Powerful primary geography is about enabling children to become informed, caring and responsible citizens in their local and global community. It is about generating a sense of awe and wonder and a lifelong fascination with the world. It can only happen by facilitating children's engagement as local, national and global citizens. Powerful primary geography involves a journey of self-discovery as much as it does finding out about the world.

> *Exercise 1.3 Personal reflection for teachers*
>
> How is geography taught in your classroom/in your school?
>
> Over the last few months, what geographical topic/theme/lesson was particularly noteworthy? Reasons for this?
>
> Are you familiar with online resources for teaching geography, e.g. the Geographical Association's website and the journal *Primary Geography*?
>
> What resources or continuing professional development (CPD) would help you to be a better geography teacher?

Further resources

Books

Barlow, A. and Whitehouse, S. (2019) *Mastering primary geography*. London: Bloomsbury.
Catling, S. and Willy, T. (2018) *Understanding and teaching primary geography*. London: Sage.
Pike, S. (2015) *Learning primary geography: Ideas and inspiration from classrooms*. London: Routledge.
Scoffham, S. ed., (2016) *Teaching geography creatively* (2nd edn). London: Routledge.
Scoffham, S. and Owens, P. (2017) *Bloomsbury curriculum basics: Teaching primary geography*. Bloomsbury Publishing.
Willy, T. ed., (2019) *Leading primary geography: The essential handbook for all teachers*. Sheffield: The Geographical Association.

Geographical Association www.geography.org.uk

The Geographical Association is the leading subject association for teachers of geography. Its mission is to 'further geographical knowledge and understanding through education'. The GA attributes significant weight and importance to all stages of learning in geography from the early years through to initial teacher education and to the continuing professional development of teachers. Its substantial membership of practitioners and professionals in each of these sectors is driven by a belief that geographical education enriches the lives of all children and young people.

A thematic unit on Bridges for Key Stage 2 is available from Council for the Curriculum Examinations and Assessment (CCEA)

http://ccea.org.uk/sites/default/files/docs/curriculum/connected_learning/thematic_units/stem/tu_bridges.pdf

Follow the things www.follow the things.com

Follow the things.com is a fake shopping website where relationships between the production, transportation and distribution of commodities have been researched. This work is published online via social media such as Facebook, Twitter, Flickr and a WordPress blog. The website has the look and feel of an online store with grocery, fashion, electronics and other departments. The website publishes research about the geography of different products, for example, T-shirts, computers and books. The research gives a rich geographical account of the journey of products alongside a critique of other realities which shape the production and distribution of the goods we purchase.

The Story of Stuff http://storyofstuff.org

The Story of Stuff is a 20-minute, fast-paced, fact-filled look at the underside of our production and consumption patterns. *The Story of Stuff* exposes the connections between a huge number of environmental and social issues, and urges us to create a more sustainable and just world.

Sourcemap www.sourcemap.com

Sourcemap is an open-source, interactive database for tracking the origins and impacts of anything from a Mars Bar to a MacBook. Sourcemap lets users create, edit and browse maps detailing the supply chain and carbon footprint of a variety of products. Anyone can create a map for just about anything imaginable and, as a socially driven site, other users can edit and add to that map, connecting the dots of where materials come from and their carbon cost. Sourcemap's geopolitical information reveals the interconnected nature of modern global culture.

Articles published in the Geographical Association's journal *Primary Geography*

Alison, J. (2006) Enlivening, re-instating and celebrating geography. *Primary Geography*, 61, 30–31.
Bell, D. (2005) The value and importance of geography. *Primary Geography*, 56, 4–5.
Bonnett, A. (2013) Geography: The world's big idea. *Primary Geography*, 82, 7–8.
Cook, I., Motamedi, M. and Williams, A. (2006) Stuff geography. *Primary Geography*, 61, 38–39.
Iwaskow, L. (2006) Learning to make a difference. *Primary Geography*, 61, 22–24.
Jackson, P. (2016) Geographies of connection. *Primary Geography*, 91, 8–9.
Parkinson, A. (2016) Digital connections in the primary classroom. *Primary Geography*, 91, 20–21.
Rawlinson, S. (2016) It's all a matter of connections. *Primary Geography*, 91, 5–6.
Richardson, E. (2009) Celebrating geography. *Primary Geography*, 68, 18–20.
Scoffham, S. (2016) Debating connections. *Primary Geography*, 91, 28–29.
Willy, T. and Sahi, I. (2016) The value of personal international connections. *Primary Geography*, 91, 32–34.

References

Barnes, J. and Scoffham, S. (2017) The humanities in English primary schools: Struggling to service. *Education 3–13*, 45(3), 298–308.
Biesta, G. (2015) The duty to resist: Redefining the basics for today's schools. *RoSE–Research on Steiner Education*, 6, 1–11.
Bonnett, A. (2008) *What is geography?* London: Sage.
Catling, S. (2011) Children's geographies in the primary school. *Geography, education and the future*, London: Continuum International, 15–29.
Catling, S. (2014) Giving younger children voice in primary geography: empowering pedagogy – a personal perspective. *International Research in Geographical and Environmental Education*, 23(4), 350–372.
Catling, S. (2017) High quality in primary humanities: Insights from the UK's school inspectorates. *Education 3–13*, 45(3), 354–364.
Catling, S. (2018) *The debate we are not yet having: What do we mean by knowledge in primary geography? Some preliminary thoughts* Paper delivered at the Charney Manor Primary Geography Research Conference, 23rd–25th February 2016, Why Primary Geography? (Unpublished).
Catling, S. and Martin, F. (2011) Contesting powerful knowledge: The primary geography curriculum as an articulation between academic and children's (ethno-) geographies. *Curriculum Journal*, 22(3), 317–335.

Colvin, R.L. and Edwards, V. (2018) Teaching for global competence in a rapidly changing world. *OECD Publishing*. https://asiasociety.org/sites/default/files/inline-files/teaching-for-global-competence-in-a-rapidly-changing-world-edu.pdf

Cook, I. (2004) Follow the thing: Papaya. *Antipode, 36*(4), 642–664.

Cook, I. et al. (2017) From 'follow the thing: papaya' to followthethings.com. *Journal of Consumer Ethics, 1*(1), 22–29.

Craft, A. (2011) *Creativity and education futures: Learning in a digital age*. Stoke-on-Trent: Trentham Books.

Craft, A. (2012) Childhood in a digital age: creative challenges for educational futures. *London Review of Education, 10*(2), 173–190.

(DES/NCCA) Department of Education and Science/National Council for Curriculum and Assessment. (1999) *Primary school curriculum. Geography*. Dublin: Stationery Office. Available on www.curriculumonline.ie/getmedia/6e999e7b-556a-4266-9e30-76d98c277436/PSEC03b_Geography_Curriculum.pdf

International Commission on Education for the Twenty-First Century and Delors, J. (1996) Learning, the treasure within: Report to UNESCO of the International Commission on Education for the twenty-first century: Highlights. UNESCO Pub.

Dorling, D. and Lee, C. (2016) *Geography: Ideas in profile*. London: Profile Books.

Dorling, D. (2017) *The equality effect: Improving life for everyone*. Oxford: New Internationalist Publications Limited.

Doucet, A. and Evers, J. (2018) Introduction. In Doucet, A., Evers, J., Guerra, E., Lopez, N., Soskil, M. and Timmers, K. eds., *Teaching in the fourth industrial revolution: Standing at the precipice*, London: Routledge, 1–7.

Gardner, H. (1999) *The disciplined mind: What all students should understand*. New York: Simon and Schuster.

Geographical Association. (2009) *A different view – A manifesto for the geographical association*. Sheffield: Geographical Association.

Geographical Association. (2011) The geography national curriculum GA curriculum proposals and rationale. Available on www.geography.org.uk/download/ga_gigcccurriculumproposals.pdf

Grant, S.G. (2007) High-stakes testing: How are social studies teachers responding? *Social Education, 71*(5), 250.

Heafford, M.R. (2016) *Pestalozzi: His thought and its relevance today*. London: Routledge.

Heafner, T.L. and Fitchett, P.G. (2012) National trends in elementary instruction: Exploring the role of social studies curricula. *The Social Studies, 103*(2), 67–72.

Hicks, D. (2014) *Educating for hope in troubled times*. London: IOE Press.

Huckle, J. (2019) Powerful geographical knowledge is critical knowledge underpinned by critical realism. *International Research in Geographical and Environmental Education, 28*(1), 70–84.

Independent. (2002) Sir Ron Cooke: 'We must assert the importance of geography in the curriculum'. From the president's address at the Royal Geographical Society's annual meeting. www.independent.co.uk/voices/commentators/sir-ron-cooke-we-must-assert-the-importance-of-geography-in-the-curriculum-179715.html

James, M. and Pollard, A. eds., (2012) *Principles of effective pedagogy*. Abingdon: Routledge.

Kennedy, R. and Murphy, M.H. (2017) *Information and communications technology law in Ireland*. Dublin: Clarus Press.

Lambert, D. (2009) A different view. *Geography, 94*(2), 119–125.

Lambert, D., Solem, M. and Tani, S. (2015) Achieving human potential through geography education: A capabilities approach to curriculum making in schools. *Annals of the Association of American Geographers, 105*(4), 723–735.

Leonard, A. (2010) *The story of stuff: How our obsession with stuff is trashing the planet, our communities, and our health-and a vision for change*. New York: Simon and Schuster.

Marshall, T. (2018) *Divided: Why we're living in an age of walls*. London: Elliott and Thompson Ltd.

Martin, F. (2005) Ethnogeography: A future for primary geography and primary geography research? *International Research in Geographical & Environmental Education*, 14(4), 364–371.

Martin, F. (2006) *Teaching geography in primary schools*. Cambridge: Chris Kingston.

Martin, F. (2008) Ethnogeography: Towards liberatory geography education. *Children's Geographies*, 6(4), 437–450.

Maude, A. (2016) What might powerful geographical knowledge look like? *Geography*, 101(2), 70–76.

Maude, A. (2018) Geography and powerful knowledge: A contribution to the debate. *International Research in Geographical and Environmental Education*, 27(2), 179–190.

McCaughrean, G. (2007) *Stop the train*. Oxford: Oxford University Press.

Morgan, A. (2006) Developing geographical wisdom: Postformal thinking about, and relating to, the world. *International Research in Geographical and Environmental Education*, 15(4), 336–352.

Nagel, P. (2008) Geography: The essential skill for the 21st century. *Social Education*, 72(7), 354–358.

Nesbit, E. (2017) *The railway children*. London: Virago.

Nussbaum, M. and Sen, A. eds., (1993) *The quality of life*. Oxford: University Press.

OECD. (2003) The definition and selection of key competencies executive summary. www.oecd.org/pisa/35070367.pdf

Owens, P. (2013) More than just core knowledge? A framework for effective and high-quality primary geography. *Education 3–13*, 41(4), 382–397.

Piff, P.K., Dietze, P., Feinberg, M., Stancato, D.M., and Keltner, D. (2015) Awe, the small self, and prosocial behavior. *Journal of personality and social psychology*, 108(6), 883.

Roberts, M. (2017) Geographical education is powerful if. … *Teaching Geography*, Spring, 42(1), 6–9.

Scarratt, E. and Davison, J. ed., (2012) *The media teacher's handbook*. Abingdon: Routledge.

Schleicher, A. (2018) *Preparing our youth for an inclusive and sustainable world. The OECD PISA global competence framework* OECD www.oecd.org/education/Global-competency-for-an-inclusive-world.pdf

Schwab, K. (2017) *The fourth industrial revolution*. New York: Crown Business.

Scoffham, S. (2013) Geography and creativity: Developing joyful and imaginative learners. *Education 3–13*, 41(4), 368–381.

Scoffham, S. (2016) *Spirit of place: Awakening a sense of awe and wonder*. In Primary Geography Research Conference, 26th–28th February 2016, Charney Manor, Oxfordshire. (Unpublished).

Shulman, L. (2003) *The wisdom of practice: Essays on teaching, learning and learning to teach*. Hoboken, NJ: Jossey-Bass.

Statment, N.P. (2009) Powerful and Purposeful Teaching and Learning in Elementary School Social Studies. *Social Education*, 73(5), 252–254.

Tanner, J. (2010) Geography and the emotions. In Scoffham, S. ed., *Primary geography handbook*, Sheffield: Geographical Association, 35–47.

Trilling, B. and Fadel, C. (2009) *21st Century skills: Learning for life in our times*. San Francisco: Jossey-Bass.

UNESCO. (2012) Shaping the education of tomorrow: 2012 report on the UN decade of sustainable development. Retrieved from http://unesdoc.unesco.org/images/0021/002166/216606e.pdf

World Wildlife Fund (WWF). (2018) Living planet report 2018: Aiming higher https://wwf.panda.org/knowledge_hub/all_publications/living_planet_report_2018

Young, M. (2008) *Bringing knowledge back in: From social constructivism to social realism in the sociology of education*. Abingdon: Routledge.

Young, M., Lambert, D., Roberts, C. and Roberts, M. (2014) *Knowledge and the future school: Curriculum and social justice*. New York: Bloomsbury Publishing.

2 Powerful geographical thinking
Initiating investigations and enquiry-based learning

Introduction

Geographical thinking and reasoning occurs when a child is fully engaged with geographical investigations such as conducting a traffic survey or assessing the tourism potential in their local area. Each investigation involves the use of geographical skills such as observation, mapping and making connections. The most successful investigations and enquiries are those of interest to children, shaped by children and framed by children's questions.

The development of thinking and problem-solving skills is essential, considering children will have to deal with future geographical problems that we cannot conceive right now. Brooks et al. (2017: 3) argue that 'geographical education is well positioned to speak directly about many current and future issues and can provide powerful ways of thinking about them'.

This chapter sets out to:

- provide an overview of geographical thinking;
- highlight the importance of asking good geographical questions;
- illustrate the process for conducting geographical investigations and enquiries; and
- provide examples of classroom work which promotes children's geographical thinking.

Geographical thinking

Thinking geographically is not the same as everyday thinking. Geographical thinking involves making connections between local and global, human and physical, near and far, time and distance, and so on. Jackson (2006: 199) says 'thinking geographically offers a uniquely powerful way of seeing the world and making connections between scales, from the global to the local'. Geographers make sense of the world by synthesising information from various sources through the use of geographical skills and concepts. Brooks et al. (2017) suggest that geographical education can provide powerful ways of thinking about contemporary and future issues. Simply knowing the capital of France or the location of the river Thames or being able to recite the states of America does not in itself constitute geographical thinking. Traditional rote-learning approaches do little to promote geographical thinking. Edelson et al. (2013: 10) advise against the teaching of geography as 'a litany of locations', arguing that the 'where' comprises the basic alphabet of geography, while the question 'why there' focuses on the connections between places and promotes sophisticated geographical thinking.

Enquiry-based learning and geographical investigations which involve decision-making and/or problem-solving bring children through a process which generates geographical thinking. The Road Map for 21st Century Geography Education Project (Edelson et al., 2013: 10) argues that essential geographic practices for promoting geographical thinking are:

- formulating geographic questions;
- acquiring, organising and analysing geographic information; and
- explaining and communicating geographic patterns and processes.

Advocates of holistic geography suggest that young people need to move beyond studying geographical topics in isolation (Reimers et al., 2016). Instead they need to think geographically by seeing the connections between different systems (Bonnett, 2012). However, some schools and textbooks employ a superficial approach to geography. Hence, there is an over-reliance on studying one topic at a time with no broader context. This exacerbates the public perception of geography as that of a 'fact-based rather than conceptual discipline', otherwise known as a 'Trivial Pursuits' version of geography (Jackson, 2006).

Lambert (2011) differentiates between geography's 'core knowledge' – the vocabulary of geography (facts and figures) – and geography's conceptual knowledge – the grammar of geography (concepts and theories which help us make sense of factual information). While both are important there tends to be an overemphasis on the vocabulary rather than on the grammar of geography. Primary geography teachers need to address both the grammar and vocabulary of primary geography.

Geographical concepts or big ideas provide a means of organising geographical topics and content. They link pieces of information together into a coherent picture through the process of geographical thinking. Geographical concepts include space, people, place, environment and scale as illustrated in Figure 2.1 and Table 2.1. These big ideas can be used to formulate questions, frame an enquiry and inform decision-making.

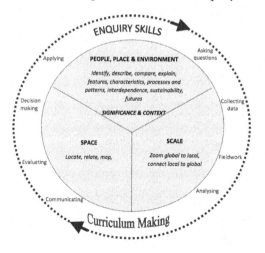

Figure 2.1 Thinking geographically
Source: GA, 2013

Table 2.1 Thinking geographically: A process that involves asking questions within a relational or holistic framework

Concept	Possible question	Elaboration
Space (locational framework)	Where is this place?	E.g. using geographical vocabulary, references, points and grids to explain in words, images, numbers, diagrams, maps and models where a place is in relation to other features and places in the world
	Why is it located here and not there?	E.g. Why are most cities located near coasts?
	How does this place connect to … ?	E.g. How easily can I get to this place and how would I travel? Why do we import goods from here and not there? Why do people move to other places? Where does our pollution go to and why?
	How does this place relate to other places?	E.g. How do nested hierarchies work?
	What is remarkable about this location?	E.g. What is it about location that makes Killarney, Co. Kerry such a successful tourist town?
	What do maps tell us about this location?	How can we use maps to interpret and talk about location? What do different maps tell us? How accurate are they?
	How can we map this place?	How can we use maps and other graphicacy tools to represent this location and information about it?
People, place, environment	What is this place/environment called?	E.g. Core knowledge of names of towns, cities, villages, types of environment, etc.
	What is it like?	E.g. in terms of characteristics, human and physical features, processes and patterns
	How did it get like this?	E.g. recognising and understanding how places are shaped by a combination of human and physical processes; knowing what those processes are and how they work
	Why is it changing?	E.g. using knowledge of human and physical processes to explain change
	What will it be like in the future?	E.g. applying geographical knowledge to imagine possible, preferable and probable future scenarios
	What do people do here?	E.g. investigating the kinds of jobs that people do, the leisure opportunities they engage with and how the landscape both influences and is influenced by such activities
	Why do people come here and what for?	
	How and why do we improve, sustain and spoil places and environments?	E.g. investigating human and physical processes and natural resources and how these influence people, environments and economies
	What's it got to do with me?	E.g. making connections between what people do and what places and environments are like
		E.g. recognising human agency
Scale	How do places change when viewed at different scales?	E.g. What happens when we 'zoom' in and out to places? What different patterns and processes do we notice?
	How do we see the global in the local and the local in the global?	E.g. How does the food I buy in my local supermarket connect me to the wider world?

Source: GA, 2013

The Thinking Geographically framework in Figure 2.1 includes those parts of Bloom's Taxonomy which develop higher-order thinking, namely applying, analysing and evaluating. The framework also includes a dimension of curriculum making as part of the process of promoting geographical thinking. Curriculum making (GA, 2013) is the creation of purposeful educational experiences which draw upon three pillars, namely: the teacher's knowledge and skills; the experiences and geographical knowledge of children; and the discipline of geography. This model considers curriculum making in terms of a process whereby the teacher creates a learning environment, geographical aims and lessons which are purposeful, meaningful, effective and enjoyable. While there is a national curriculum in place and there is pressure to deliver, there is still teacher autonomy. There are a number of factors which influence teachers' abilities to be curriculum makers including teacher confidence, geographical knowledge, ethos within the school, support from the principal, and willingness to take a risk.

Young children are divergent open thinkers. Their questions express a sense of curiosity and excitement about the world. As teenagers the impetus to ask questions is less obvious. The challenge for the teacher is to promote, reward and acknowledge good questions. Teaching children how to ask good questions takes time. An initial starting point for any enquiry may be the generation of several questions. Questions can be grouped and categorised as a means of organising a framework for thematic investigation. Through a process of negotiation and discussion, big questions, key questions and enquiry emerge.

Young children have difficulty with the nested hierarchy relationship as they tend to see Galway, Ireland and Europe as separate places. Russian dolls, also known as stacking dolls or nesting dolls, are useful for teaching the idea of 'nested hierarchy', i.e. one place is located within another. As children and student teachers reflect on 'my geography' they begin to analyse the relationships between place and space in the context of their own lived experiences. When children find out about places at a range of scales using Google Earth and a variety of maps they can further conceptualise the idea of 'a nested hierarchy'. There are many ways of illustrating the concept of nested hierarchies. Barlow and Whitehouse (2019) use a paper plate to illustrate how different elements of one's address can be drawn in concentric circles and cut up in a spiral (Figure 2.2). Similarly, the elements of a child's address can be written on separate pieces of paper to be arranged in the correct order.

Figure 2.2 Illustration of nested hierarchies

Metacognition or 'thinking about thinking' has been defined by the OECD as 'a second or higher-order thinking process which involves active control over cognitive processes' (Mevarech and Kramarski, 2014: 36). Researchers suggest that 'metacognitive approaches to teaching and learning have the potential to radically improve the outcomes and life chances of children' (Perry et al., 2018: 14). Metacognition skills are located within the broader group of 21st-century skills (Van Laar et al., 2017); and the literature suggests that it is important to teach them effectively. However, the commitment to teach thinking skills and metacognition does not feature strongly in curriculum documents and high-stakes testing. While metacognition is cross-curricular, the term *geographical metacognition* recognises the potential of geographical enquiry-based learning to develop children's thinking skills.

Asking questions

A seemingly simple event can raise a host of complex geographical questions. Enquiry-based learning is based on asking questions. Children ask questions to construct meaning and make sense of the world around them. While cross-curricular in nature, questioning is a key geographical skill which helps children make links and connections. Questions enable children to understand new concepts in the context of their previous experiences. The importance of questions in geographical enquiry is widely acknowledged (Roberts, 2013). According to Pickford et al. (2013: 87) 'good enquiry questions spark the children's curiosity and sense of wonder, because the questions usually stem from something which really matters to the children themselves and which they want to know more about and understand'. Learning how to pose questions is the first step into geographical investigation and enquiry. When given support, space and time to frame their own questions, children become interested and curious; in other words they are more interested in having their own questions answered.

Enquiry is all about arousing curiosity through asking questions. A sense of mystery and intrigue can be generated in the class as children set about the mission of enquiry which they have devised. Personal commitment by the children is generally greater when they have generated the questions. Devising good questions is one of the crucial features of geographical investigations. Some initial questions include the following:

- What is it?
- Where is it?
- What is it like?
- How did it come to be like this?
- How is it changing and what might happen next?
- How do I feel about this issue *or* What would it feel like to be in this place?

There are many frameworks and approaches which can be adopted for asking questions. The key questions of geography (Table 2.2) provide a good starting point:

By building investigations based on children's questions a greater level of personal interest is established. Kidman (2018) found that there was a mismatch between topics of interest to children and those taught by teachers. She suggests that a better fit between curriculum and children's interests could lead to improvements in cognitive outcomes and general interest in geography.

Table 2.2 Key questions of geography

Key question	Concept	Language skill
What is it?	Identification and naming of features and geographical processes	Development of specific geographical vocabulary relating to human, natural and environmental processes. This can occur through primary sources such as the local environment or secondary sources such as stories, photographs and newspaper articles.
Where is it?	Location of places on a variety of maps	Describing the location of places starting off with prepositions such as beside, inside, underneath, around, outside; left and right; and specialised geographical language including coordinates, latitude and longitude and compass directions. Articulating directions provides children with an opportunity to practise this kind of language.
What is it like?	Predicting, imagining and development of different scenarios	Use of descriptive language including various adjectives, adverb and nouns.
How did it come to be like this?	Explaining, understanding and comprehension	Explaining an idea in different ways and from different perspectives. Sequencing, grouping and ordering of ideas and key questions. Suggesting, developing and exploring alternative explanations.
How is it changing and what might happen next?		Proposing ideas about places and the processes of change. This can also include children's feelings about a place or an issue. Persuasive language also features when children are arguing from a particular perspective, e.g. in the context of land development, children may argue from the perspective of the developers or the environmentalist.
How do I feel about this issue? or What would it feel like to be in that place?	Evaluating, expressing opinion and caring	Language dealing with questions of feelings, imagination and speculation draw upon expressive language. Responding to places can evoke aesthetic and emotional responses which can be captured in descriptive language through poetry, speeches, art and descriptive language.

Source: adapted from Scoffham (2010)

Case study 2.1 Use of mascots for generating enquiry questions (4–7 years)

Many classes have their own mascot usually in the form of a teddy, puppet or cuddly toy. Children can take it in turns to take their teddy when they go away for a weekend or on holiday. They can take photos of their teddy in different geographical locations undertaking different activities: on the plane; at the beach; sightseeing; and beside local landmarks. These photographs provide a scaffold for children for talking about their geographical adventures. Having a class mascot adds a little fun and humour to the classroom. Class mascots can help the children on their learning journey. Some teachers write a class blog from the perspective of their mascot. Mascots when used regularly develop their own personality and some even have their own annual birthday parties. For younger children, the mascot becomes a big part of the classroom community and an opportunity for facilitating children to share their imaginative ideas and theories.

40   *Powerful geographical thinking*

Barnaby Bear, a character developed by the Geographical Association, has supported children's learning about other places through the enquiry question: Where in the world is Barnaby Bear? Barnaby is a teddy bear who travels the world, meets new people, has exciting geographical adventures and learns more about the places he visits. Young children love to see their familiar teddy bear in unfamiliar places and this helps to bridge the gap between their place and faraway places. Appearing as a puppet and/or a teddy bear for use in the classroom, he is accompanied with supplementary material, worksheets, DVDs, a BBC series, and different outfits. The official Barnaby Bear website www.barnabybear.co.uk includes games and resources, case studies demonstrating how several teachers have used Barnaby in their classrooms, interactive games, and a photo gallery of Barnaby on his travels.

Through Twitter a teacher from Israel made contact with a teacher in Scoil Chaitriona Junior School, Renmore, Co. Galway. Her class (5–6 years) sent a toy, some letters and some symbols from Israel. Children learnt about the Star of David, they heard some stories about Israel and they prepared their return letters. They were introduced to Barnaby Bear, who coincidentally was planning a trip to Israel. The children brainstormed a list of questions (Table 2.3) for the famous travelling bear. Photographs from Israel were also requested (Table 2.4).

Table 2.3 Questions from Senior Infants, Renmore

Geographical themes and concepts	*Geographical questions*
Location	Where is Israel?
	How will you get there?
Food and drink	What kind of food do you eat?
	Do you have McDonald's?
	Do you have carrots, potatoes, broccoli and peas?
	Do you have eggs?
	Do you have apples?
	What kinds of drinks do you have?
School	What time do you go to school?
	Which days of the week do you go to school?
Religion	What day do you go to the synagogue?
Games and leisure	Do you have *Angry Birds*?
	What games do you play?
	Do you have drawings/art?
	What movies are on?
Locations and landscape within Israel	What kinds of places do you visit in Israel?
	Do you have stone walls like us in Ireland?
Facilities and services	Do you have a cinema?
	Do you have fridges?
People	Who is the king?
Weather	Is it always hot there?
Culture and customs	Do you have chopsticks?

Powerful geographical thinking 41

Table 2.4 Photographs from Israel were requested of the items listed

Houses	Shops	Clouds
School	Hospital	Furniture
Clothes	Food	Flowers
Teachers	Tellies (TVs)	Water
Weather	Beaches	Castles

Barnaby is a wonderful ambassador for geography and children love spending time in his company. The geographical ideas behind the travelling bear have also been adopted by many classrooms using a different class mascot (Figures 2.3 and 2.4).

Critical geographical thinking

Critical thinking is a mental process of actively and skilfully conceptualising, applying, analysing, synthesising and evaluating information to reach a conclusion or solve a problem. As a core component of 21st-century skills, critical thinking means preparing children for a global society defined by high-speed communications, complex and rapid change,

Figure 2.3 Daisy's trip to Rome

42 *Powerful geographical thinking*

Figure 2.4 Children with their class mascots from St Augustine's National School, Clontuskert

and increasing multiculturalism. By thinking critically, children use multiple strategies to solve a problem. Through critical thinking, children refine their thinking process by learning how to analyse, assess and reconstruct the actual process of thinking. Children need to have the opportunity to consider differing and sometimes contrasting points of view, using different technologies and approaches. Twenty-first-century learners will join a workforce which requires problem-solvers, critical thinkers and investigators. Learners need to be able to construct knowledge, to develop and defend new ways of thinking. However, in a push for better test scores, through high-stakes testing many children are leaving the formal education system lacking the critical thinking skills that are necessary to succeed in higher education or in the workplace (Smith and Szymanski, 2013).

Ruggiero's (2012) characteristics of critical thinkers are especially pertinent for geographers. Critical thinkers are honest with themselves and are mindful of the limitations of their own knowledge and perceptions. Perpetually curious, they strive for understanding and are interested in other people's ideas and opinions. They regard problems and controversial issues as exciting and challenging. Judgements are based on evidence. Critical thinkers recognise that extreme views (whether conservative or liberal) are seldom correct, so they avoid them, practice fair-mindedness, and seek a balanced view, and they practise restraint by thinking before acting.

What is enquiry?

As adults we constantly engage in enquiry and problem-based learning. Whether it is learning how to bake a cake, make a video, bleed a radiator, we do not return to school to learn these tasks. Instead, we use everyday enquiry, which may involve asking a friend, searching the internet, looking at a YouTube clip, seeking specialist advice, trial and error, or a mixture of all the aforementioned. Problems, such as being locked out of one's house, coping with a flat tyre, dealing with a sick child (although more stressful), still require solutions. The process of dealing with these problems is not always recognised as a learning process. This form of participative, experiential informal learning deserves wider recognition.

The terms *investigation*, *problem-based learning* and *enquiry-based learning*, all associated with promoting thinking, have much in common. Each investigation generally progresses through a number of steps: asking questions, designing an investigation, collecting data, drawing conclusions and communicating findings. The term 'enquiry-based learning' is generally associated with geographical investigations. Geographical enquiry learning is based on investigations of the local, regional, national and global world by asking questions, finding solutions and presenting findings.

An enquiry approach to geographical study may be defined as one in which the teacher works with children in developing their abilities to ask geographical questions. Through investigative work the children seek to answer these questions leading to sound knowledge, understanding and skill development. Enquiry helps children to engage with the world through asking their own questions and conducting their own investigations. It helps children to engage geographically with their world through observation, questioning, research, proposing solutions and critical reflection. According to Leat (2001: 29) 'enquiry is at the heart of understanding ourselves and our relationship with the world'.

Enquiry is an approach to teaching and learning which extends geographical conceptual understanding, geographical thinking and skill development simultaneously. Through enquiry-based learning, the process of learning is as important as the core knowledge or concepts under exploration. Enquiries are also influenced by children's prior knowledge. Enquiry-based learning 'is an approach to learning disciplinary knowledge that enables students to develop a critical understanding of the world' (Roberts, 2013: 50). Pickford et al., 2013: 84) state that 'enquiry-based learning through the humanities encourages schools to focus on developing the kinds of educational experience appropriate for young people growing up in the 21st century age of globalisation'. Roberts (2013) sets out four essential aspects of geographical enquiry learning namely: creating a 'need to know' through enquiry questions; the importance of using geographical data as evidence; the process of making sense of this data; and the importance of reflecting on learning. Figure 2.5 illustrates some frameworks for enquiries developed by student teachers using Roberts' cycle.

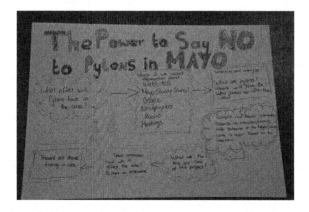

Figure 2.5 Enquiry frameworks developed by student teachers (see Section 3 of the colour plates for more examples)

44 *Powerful geographical thinking*

Enquiry is rooted in important educational principles and in particular is informed by constructivist ideas of Vygotsky (1978, 1986) and Dewey (1990, 2007). John Dewey in his book *Experience and Education* (2007: 16) described processes of rote learning as 'static', referring to traditional education as an 'imposition from above and from outside' (p. 16). He argued that rather than filling children's minds with isolated facts and figures, children should be actively involved in their learning and help to co-construct knowledge that has both interest and meaning to them. In order to do so the role of the teacher should change from that of an 'external boss or dictator' to that of a 'leader of group activities' (Dewey, 2007: 45). Dewey's work is even more pertinent in the context of 21st-century learning competencies.

Enquiry-based learning is based on constructivist approaches to teaching and learning whereby the child constructs meaning in relation to prior knowledge. The nature of teaching and learning has changed radically in recent years with a move from traditional, didactic forms of teaching to child-centred approaches whereby children actively construct knowledge about their world (Table 2.7). Active learning and hands-on experience with natural phenomena such as the weather and the local environment enhance conceptual learning and geographical thinking.

Constructivist pedagogy is rooted in the idea that knowledge is not an objective construct but emerges as we engage with and experience reality. Constructivism is a learning theory describing the process of knowledge construction. Knowledge construction is an active rather than a passive process. It emphasises the importance of the teaching context, children's prior knowledge, active engagement of the child and interaction between the learner and the emerging knowledge. In the constructivist perspective, knowledge is constructed by the child through his or her interactions with the environment. However, there is a danger that enquiry 'becomes everything and nothing'. It is important that enquiry is not reduced to 'a bit of finding out'.

The enquiry process begins with an enquiry question. This big question may arise from a series of smaller questions posed by the children. Through a process of clarification and prioritisation a final big question is agreed with the whole class. This forms the basis of the enquiry. Once the question is selected the class and the teacher decide how this question is going to be answered, what data is going to be collected and how it is going to be collected. Following data collection, it is analysed and conclusions are drawn. Generally the class has an opportunity to share its findings with a wider audience.

Case study 2.2 Children's newspaper as a model of enquiry

St Brendan's NS Eyrecourt is a co-educational school with three mainstream teachers in the village of Eyrecourt, a small village in County Galway. Eyrecourt and Eyre Square in Galway city are named after the Eyre family. John and Edward Eyre from Wiltshire in England came to Ireland in 1649, as part of the Cromwellian army that invaded Ireland. John settled in Eyrecourt on land formerly owned by the O'Madden family, while his brother Edward based himself in Galway city. Once a major urban settlement, Eyrecourt now faces the challenges experienced by many towns and villages in rural Ireland including rural depopulation, closure of businesses and lack of employment. Since I visited the school, the local post office has closed down as part of a national re-organisation of the postal services. This illustrates the dynamic and ever-changing essence of local geography. Nevertheless Eyrecourt is a dynamic community hub with a strong sporting identity and a fascinating history.

Figure 2.6 Children involved in producing their own newspaper

Senior classes (11–13 years) from St Brendan's School publish their own newspaper called *The Eyrecourt Examiner* (Figure 2.6). Each year an editor and a group of sub-editors are selected through a written application process. The sub-editors are assigned to the following portfolios: sport, agriculture and entertainment. Working as editors and sub-editors the children research, write and design their newspaper. The chief editor and sub-editors have regular newspaper meetings with the class. Children volunteer to write stories and cover particular events. The newspaper covers school events, local and national events, sporting matches, concerts, day trips and films. Many children write about places they have visited with their families, e.g. Ailwee Caves, Eagle Sanctuary and the 3 Arena in Dublin. Local links to international events are also covered. The children's newspaper featured in the national newspaper *The Irish Times* as an example of innovative teaching and learning. The children have also presented their work to teachers at a national computer education conference and to student teachers in Mary Immaculate College, Limerick. It is a very good example of child-directed learning skilfully facilitated by the classroom teacher. Available for sale in local newsagents, the newspaper is sold to the public and the children decide how the profits will be used after the publication costs are deducted.

Lucidpress is the publishing software used. Newspaper articles are saved on Google Drive and all children are able to upload their articles for editing. The children regularly look at other newspapers and their online presence. When a child types up an article he or she also has to do a taster summary of the story for the newspaper Weebly account. It is saved as a draft and, in school, it is checked and published by the teacher. This account is linked to the newspaper's Twitter account @Eyrecourt_News.

The school has a digital microphone recorder which is used when interviewing visitors. Each child who conducts and records an interview is responsible for its final written version in the newspaper. After a final edit under the watchful eye of the teacher, the newspaper goes to print! With a colour photocopier in their school, the children print 100 copies of the *Eyrecourt Examiner*. Sold in Eyrecourt village (courtesy of the two shops, pharmacy, and the hairdresser's), the children decide how the profits are spent. To date the profits have funded several trips and some of it has been re-invested in camera and iPads for the journalists and researchers.

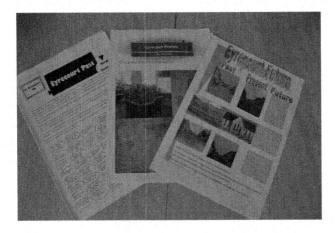

Figure 2.7 Newspaper supplements published by the children: Eyrecourt Past, Present and Future

Figure 2.8 Children and principal teacher from St Brendan's NS Eyrecourt presenting their newspaper to the President of Mary Immaculate College, Limerick

Initiating research for local study in Eyrecourt, Co. Galway

As part of the research for this book the children produced three supplements: *Eyrecourt Past, Eyrecourt Present* and *Eyrecourt Future* (Figure 2.7). The *Eyrecourt Past* supplement is still on sale in the village as a standalone copy for tourists. Enquiry questions from the children informed the first steps for their research (Figure 2.9). Two enquiry or big questions emerged from discussions:

1   What is the story behind the name Eyrecourt?
2   Once a significant cosmopolitan urban setting, why is Eyrecourt a quiet village today?

Powerful geographical thinking 47

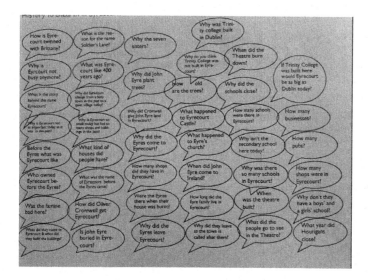

Figure 2.9 Children's enquiry questions

Feedback from the children

The children reflected on the nature of their work for the newspaper supplements. A summary of their comments is provided in Table 2.5.

Table 2.5 Children's reflections about the supplements

Supplement: Feedback from children

I thought it was hard work to put it together. I learnt a lot because I did not know a lot of the information.

The supplement was very good because we are adding on more pages to the Eyrecourt Examiner and if someone read the paper and knew nothing about Eyrecourt they would learn a lot. I thought it was much better than reading out of a book.

It has changed the way we think about Eyrecourt's history and geography.

There were a lot of schools in Eyrecourt. The library used to be a school. There was a secondary school at the back of Roger Whittaker's house.

I would have liked to see and know about the school down River Street long ago.

I thought the process was very good and I found out a lot about Eyrecourt and can't wait to do the next supplement. It was a great enjoyable experience.

It was great telling everybody about what we learnt about Eyrecourt. It was fun to think what it was like in Eyrecourt long ago.

The activities helped us to learn more about Eyrecourt. Now we notice all the arches. We liked it because it was something different. We're proud of this supplement.

I thought it was fun because we were able to do maps and geography walks. We generated a lot more information than we thought we would at the beginning of the project. My parents loved it.

48 *Powerful geographical thinking*

Table 2.6 Feedback from parents

I learnt a lot about the history of the Eyres and how Eyrecourt got its name. I loved the articles on Eyrecourt Theatre, John Eyre of Eyrecourt and Eyrecourt Castle. I thought they were very informative.

The images are great, maybe an appeal to residents of Eyrecourt might bring in more. The names of the shops/businesses in the past was interesting. I learnt there were a lot of business in Eyrecourt in the past.

I enjoyed the story of Eyrecourt in the past. The comparison between Eyrecourt in the past and present day Eyrecourt was interesting.

It was a very good piece of journalism. I am not from Eyrecourt originally. I found the article about Eyrecourt's history very interesting and also well researched. I will keep the supplement as a keepsake for future visitors.

Didn't realise there was as many businesses and residents in Eyrecourt in the past. There was nearly 2,000 people living in Eyrecourt in the 1800s. Today there are less than 300 people here.

Suggestions for future articles

Maybe focus on one particular building or area per issue. I would have liked to have seen some images of local houses such as Prospect House and Hagney's house.

Wildlife in the area would be an interesting topic. Another idea would be to include surrounding areas such as Meelick and Clonfert.

Include some information about emigration and why so many people had to leave Eyrecourt. What about an interview with senior citizens?

Perhaps you could cover some history on the school when it was run by the nuns, the old boys' school and the old convent. You could do some research on famous people from Eyrecourt.

Please look at the decline of Eyrecourt and the lack of facilities here. Include an article about the absence of broadband.

I would like to see an article about the Garda station in the past along with the jail and the courthouse. Suggestions for the renewal of the village. Or is the village irrelevant today; explore pros and cons.

Suggestions for doing up some of the old buildings or ideas for Eyrecourt in the future.

I don't think it could be improved as it is an excellent supplement. Keep up this important work. Well done!

The children actively sought feedback from parents and a summary of their feedback is provided in Table 2.6.

Working as reporters researching climate change

Following a workshop on Climate Justice facilitated by Trócaire, a development non-governmental association, the children decided to compile a supplement on the issue to spread awareness with the wider community. During the workshop they learnt about fossil fuels and the carbon cycle. They discussed the impact humans have on levels of greenhouse gases and they assessed different options for action. The children compiled a supplement for their newspaper based on the enquiry question: *Is it fair that our pollution affects their lives?* For their supplement they researched greenhouse gas emissions in their local area, they compared agriculture in Ethiopia with their local area and they made recommendations for action. The education officer from Trócaire was interviewed by the children (Figure 2.10). This work deepened the children's understanding of climate change and the supplement symbolised their commitment to working as climate justice advocates (Figure 2.11). Their supplement was brought to Ethiopia by Trócaire as an example of work conducted by Irish children.

Powerful geographical thinking 49

Figure 2.10 Children investigating climate change

Figure 2.11 Newspaper supplement on climate change

Case study 2.3 Sacred Heart Primary School, Granard, Co. Longford

As part of the Junior Entrepreneur Programme (JEP) 6th class children (11–13 years) from Sacred Heart Primary School, Granard, Co. Longford researched, wrote and illustrated their own book about their local area. Entitled *Granapedia: The first ever encyclopaedia of Granard: The A to Z of Granard*, this book is an in-depth exploration of the geography and history of the children's local place (Figure 2.12). The book is for sale in the local area at €5. To date 250 copies have been published.

50 *Powerful geographical thinking*

Figure 2.12 An extract from the first ever encyclopaedia of Granard: The A to Z of Granard

The Junior Entrepreneur Programme (JEP) is an entrepreneurial awareness and skills enhancement programme for primary school children. The programme aims to help children recognise enterprise and entrepreneurship and to foster an awareness and understanding of the entrepreneur's role in the community, therefore empowering the child to start to think and act with the initiative, creativity and independence that are invaluable in the modern world. As part of JEP, primary school children get a chance to experience involvement in business. The children create a product which they try to sell in their school and community.

JEP is a ten-week programme. For the first one to three weeks, the children split into groups and create their ideas as a group for the product which is to be sold. Each group develops an innovative business idea. Then each group presents their project to a team of 'Dragons' who put them through their paces and after some deliberation choose the project which they feel has the most potential as a business proposition. The next seven weeks are spent researching, designing and producing the final product.

The programme is a very positive initiative for the children involved, helping to build confidence, self-awareness and self-esteem, and an appreciation of each individual's skills and talents. This innovative programme has opened the minds of the children to the joy of entrepreneurship at a time when they are full of imagination and open to new possibilities. The programme helps the children to develop a number of skills including literacy, writing, presentation, drawing, technology, research skills, numeracy, financial skills, consumer awareness, storytelling, listening skills, creative thinking, problem-solving, team-building and collaboration skills. JEP complements the school curriculum, particularly in areas such as maths and science, but also helps children develop skills in presentation, drawing and collaboration.

The changing roles of teachers in developing enquiries

The case studies discussed in this chapter showcase inspiring enquiry-based, child-led geographical investigations. The nature of enquiry-based learning illustrated in these case studies suggests different and emerging roles for teachers and children. The role of the teacher is crucial. Teachers need to be comfortable working within enquiry frameworks. Both subject and pedagogical knowledge are prerequisites for geographical enquiry. Harvey

Table 2.7 Significant differences between traditional didactic pedagogies and enquiry-based learning.

Traditional didactic teaching	Enquiry approaches
Teacher selection and direction	Children's voice and choice
Assigned topics and isolated facts commonly known as the 'rivers and mountains list'	Questions and concepts
Working individually	Working collaboratively
Learning facts off by heart	Strategic thinking whereby children make connections and base their opinions on previous experiences
Use of textbook case studies	Authentic real-world investigations
Child compliance	Child responsibility
Child as information receiver	Child as knowledge creator
Teacher as expert	Teacher as facilitator, coach and model
Reliance on a textbook	Use of several resources
Reliance on written sources	Multimodal learning
Child hears about geographical activities	Child engages in geographical learning
Extrinsic motivators, e.g. test results	Real purpose and audience
Forgetting and moving on to the next topic	Caring and taking action
Ticking boxes and filling in blanks	Performance and self-assessment

Source: adapted from Harvey and Daniels (2009)

and Daniels (2009) set out a number of significant differences between traditional didactic pedagogies and enquiry-based learning and Table 2.7 is an adaptation of their ideas.

The teacher's role in enquiry is crucial. At the planning stage the teacher's expert knowledge guides the children in their formation of key questions. Using strategies which facilitate interactive teaching approaches, child-generated questions and ideas provide the interest and motivation for learning. The role of the teacher in enquiry is influenced by Vygotsky's (1962) ideas about scaffolding. Vygotsky's Zone of Proximal Development (ZPD) refers to the gap between what children can already do and what they could do with some limited assistance. The implication for 'good' geography lessons is, first, that teachers need to be aware of each child's starting point and, second, that they need to plan how to support progress within each child's ZPD. Through scaffolding teachers can work with children to identify goals and solve problems.

Exercise 2.1 Personal reflection for teachers

Think about one recent positive learning experience in your life (e.g. learning a new skill or overcoming a particular challenge).

- What were the factors which contributed to this learning?
- How and why were you motivated to learn?
- What was your role in this learning?
- Was there another person or resource involved?
- Did an enquiry process have any role in this learning experience?

Using mysteries for promoting geographical thinking

Geography is essentially about relationships between people, places and the environment. These relationships can be witnessed through numerous physical, natural and human processes such as erosion, earthquakes and urban development. Some of these processes are visible while others are more subtle. A mystery begins with a challenging question that invites children to investigate an issue and solve the problem posed. An effective way to promote geographic thinking, mysteries are particularly good for promoting understanding of causes, processes and consequences (Leat, 2001). To solve the mystery in groups, children receive 15–30 pieces of paper. According to Karkdijk et al. (2013) the key features of mysteries are that:

1. they are built around natural or human phenomenon with a great impact, such as a tsunami, landslide, factory shutdown or climate change;
2. the challenging question has to do with the real-life situation of someone in order to enhance children's imagination of the issues at stake;
3. there are different correct ways to answer the question;
4. children are provided with diverse and unconnected information. Information can be specific and detailed or general and abstract. So-called 'red herrings' might also be included, information that seems to be useful but is not necessary or is even misleading;
5. collaboratively children rearrange the 15–30 pieces of information, sorting them into groups or categories. Based on this, children can suggest a solution to the mystery giving reasons for their choice;
6. all solutions are discussed either in groups or with the whole class. The children also have an opportunity to discuss any other points which were raised during the exercise and their reflections on what they learnt.

The most important part of the whole process is the debriefing which takes place afterwards. This allows children to reflect on the process, the suggested solutions and the reasons for these solutions. An example of a geography mystery which deals with the topic of plastic, entitled *Where did Orla's Barbie end up?*, is set out below (Table 2.8, Figure 2.13).

Table 2.8 Mystery: Where did Orla's Barbie end up?

A million plastic bottles are purchased every minute across the world.	China has been the largest global importer of many types of recyclable materials. In one year Chinese manufacturers imported 7.3 million metric tonnes of waste plastics from countries including Ireland, the UK, the USA and Japan, and from other European Union countries.	China has now banned imports of 24 categories of recyclables and solid waste. This campaign against *yang laji*, or 'foreign garbage', applies to plastic, textiles and mixed paper.
Orla notices her recycling bin has large quantities of plastic every week.	Orla and her brother went fishing in the river Suck, Ballinasloe. They were surprised to see so much plastic floating in the river.	Almost every plastic bag ever made is still on the planet in some form or another.

Plastic is everywhere. In fact, most people would agree plastic has become an almost unavoidable part of modern life.

Many children's toys are packaged inside oversized boxes and wrapped in layer upon layer of plastic wrap.

Some harmful plastics can leak chemicals into our food and drinks, and into the environment.

80% of all toys distributed worldwide are made in China.

Most toys are made from petroleum-based plastic. Toy manufacturing has a huge impact on our health and the environment.

99% of the plastic that should be floating in the oceans is missing. Millions of tonnes have simply disappeared. As most plastic never deteriorates, it simply breaks down into smaller and smaller particles that are invisible to the human eye; what happens to this missing ocean plastic is a mystery. Scientists suggest the plastic is home to a new ecosystem: the plastisphere.

Plastic is a wonderful resource because it is so durable. It is also a terrible resource because it is so durable.

Orla donated her old toys to the local charity shop.

Charity shops sell second-hand materials including toys.

Orla's favourite toy was her Barbie doll. She had this doll since she was three and she played with her faithfully throughout her childhood. As it is made of plastic it has lasted well. It is in good condition but Barbie's hair has seen better days.

Betty's mother tries to buy toys in the local second-hand shop but Betty wants to go to Toys R Us or Smyths Toys Superstore because her friends buy toys there.

Orla's school organised a bring 'n' buy sale of old toys and books, the proceeds of which went to support a local charity.

According to scientists, by 2050 there will be as much plastic by weight in the ocean as fish.

The world produces more than 300 million tonnes of plastic every year including billions of plastic bottles and plastic bags. Half of this is used just once and then thrown away.

Only a fraction of the plastic we produce is recycled.

Mr Concannon said his company was willing to take Ireland's used plastic bottles and convert them into raw materials for pipes which could be used for roads, housing and other drainage purposes. However, local authorities would have to establish more communal collection services. This could include an additional bin for collecting plastics.

Based in the West of Ireland, JFC is an international business renowned for manufacturing innovative, high-quality *plastic* products. It provides much-needed jobs in the local area.

Galway firm JFC Manufacturing has become the largest recycler of plastic bottles in Britain following two acquisitions in England. JFC Manufacturing has bought Delleve Plastics in Stratford-upon-Avon and Reprise Plastic Recycling in St Helens. The two firms convert plastic bottles mainly to corrugated plastic pipes.

(Continued)

54 *Powerful geographical thinking*

Table 2.8 (Continued)

| Orla uses a stainless-steel water bottle which she brings to school every day. | A million plastic bottles are bought every minute around the world and they make up a third of the plastic litter in the seas. | One noticeable success in Ireland's environmental track record was the introduction of a plastic bag levy in 2002, the first country in the world to do so. The National Litter Pollution Monitoring System showed that when the levy was introduced, 5% of all litter was plastic bags. |

Figure 2.13 Where did Orla's Barbie end up?

Conclusion

Thinking geographically involves children working as geographers conducting enquiries and investigations. An important 21st-century competency, learning to think helps develop reasoning and analysis, and helps children to make learning connections. Enquiry-based learning occurs when children ask questions, conduct research, complete investigations and present their findings. It generates intensive powerful geographical thinking. Decision-making and problem-solving using real-life local and global issues can promote powerful geographical thinking.

During the process of research for this book, one principal commented that the curriculum is a mile wide and an inch deep. Through the process of curriculum making, teachers and children can make the curriculum work for them. They can become active agents in selecting investigations which are of interest to their class, underpinned by the curriculum and promote deep learning. Enquiry when it is done well promotes deep learning. Engaging in an enquiry generates curiosity about the children's own experiences and about events and processes in the everyday world. Through the case studies presented in this book, I witnessed children confidently asking questions, seeking answers and developing their own ideas. Children used technology competently in their quest to create knowledge. The children were not afraid to take a risk and they demonstrated an ability to make sense of their world notwithstanding its inherent complexity. Children were acutely aware of the changing nature of knowledge and they appreciated the subjective nature of evidence, images and geographical sources. Geographical investigations and enquiry-based learning have the power to change the way children think. This surely is one of the most important 21st-century competencies required in dealing with the challenges facing contemporary society today.

Further resources: thinking and enquiry

Brooks, C., Butt, G. and Fargher, M. (2017) *The power of geographical thinking*. Dordrecht: Springer.

Eyrecourt newspaper project: More information is available from the website eyrecourtexaminer.weebly.com/ and follow us on Twitter @Eyrecourt_News.

Pickford, T., Garner, W. and Jackson, E. (2013) *Primary humanities: Learning through enquiry*. London: Sage Publications.

Roberts, M. (2013) *Geography through enquiry approaches to teaching and learning in the secondary school*. Sheffield: Geographical Association.

Articles published in the Geographical Association's journal *Primary Geography* about geographical enquiry

Bird, S. and Cooper, P. (2006) Microclimate enquiry and fieldwork. *Primary Geography*, *59*, 26–27.

Clark, K. (2010) Developing an enquiry-based approach to geography. *Primary Geography*, *72*, 10–11.

Flack, J. (2015) Konnichiwa personalising enquiry. *Primary Geography*, *87*, 28.

Green, J., Shaw Hamilton, A., Walsh, M.O., Mahony, O. and Pike, S. (2013) On our own (with a bit of help). *Primary Geography*, *80*, 28–29.

Lakhani, S. (2010) Do we care? An enquiry approach to discovering the local area. *Primary Geography*, *73*, 10.

Lane, A. (2006) Talking through enquiry. *Primary Geography*, *60*, 16–17.

Murdoch, K. (2014) The urge to enquire: Curiosity in the classroom. *Primary Geography*, *85*, 10–12.

Pike, S. (2013) Planning our curriculum making: The influence of John Dewey. *Primary Geography*, *82*, 26–28 Autumn.

Pike, S. (2017) To challenge or not? Lev Vygotsky and primary geography. *Primary Geography*, *94*, 26–27 Autumn.

Potts, R. and Martin, H. (2014) Enquiring minds and unique places. *Primary Geography*, *83*, 26–27.

Richardson, P. (2006) The importance of speaking and listening in enquiry fieldwork. *Primary Geography*, *59*, 9–10.

Short, K. (2012) Engaging pupils in dialogue. *Primary Geography*, *78*, 6–7.

Stanton, L. (2005) Enquiry in the early years. *Primary Geography*, *56*, 22–24.

Swift, D. (2013) Thinking it through. *Primary Geography*, *80*, 7–8.

Whitehouse, S. and Vickers-Hulse, K. (2016) Generating enquiry skills. *Primary Geography*, *90*, 6–7.

Whiteley, M. (2015) Connecting with Asia though enquiry. *Primary Geography*, *87*, 29.

Wood, S. (2013) Progression in questioning skills. *Primary Geography*, *80*, 22–23.

Articles about geographical thinking

Barlow, C. (2017) Geography of the imagination. *Primary Geography*, *92*, 14–15.
Bowden, R. and Swainston, H. (2012) Thinking and talking together. *Primary Geography*, *78*, 12–13.
Catling, S. (2009) Thinking of Britain in children's geographies. *Primary Geography*, *69*, 16–19.
Catling, S. and Taylor, L. (2006) Thinking about geographical significance. *Primary Geography*, *61*, 35–37.
Chorekdijan, L. (2017) Critical thinking to promote sustainability. *Primary Geography*, *93*(2), Summer 24–25.
Ellis, S. and McCarthy, M. (2017) Challenging geography through silent debate. *Primary Geography*, *80*, Autumn 7–9.
Greenwood, R. (2017) A woodland 'what if … ?'. *Primary Geography*, *80*, 22–24.
Hopkins, J. and Owens, P. (2016) Critical thinking in geography. *Primary Geography*, *91*, 23.
Lewis, R. and pupils at Kingscourt School (2013) Challenging thinking about food miles. *Primary Geography*, *81*, 10–12.
Riley, J. and Cook, J. (2012) Discussion heads: Understanding-thinking. *Primary Geography*, *78*(2), Summer 24–25.
Shah, H. (2010) Critical thinking in the context of global learning. *Primary Geography*, *71*, 14–15.
Spina, A. (2015) Thinking and learning like a child. *Primary Geography*, *86*, 24–25.
Swift, D. (2005) Valuing places: Thinking geographically. *Primary Geography*, *58*, 4–7.
Swift, D. (2013) Thinking it through. *Primary Geography*, *80*, 7–8.

References

Barlow, A. and Whitehouse, S. (2019) *Mastering primary geography*. London: Bloomsbury Publishing.
Bonnett, A. (2012) Geography: What's the big idea?. *Geography*, *97*(1), 39–41.
Brooks, C., Butt, G. and Fargher, M. (2017) Introduction: Why is it timely to (re) consider what makes geographical thinking powerful. In Brooks, C., Butt, G and Fargher, M eds., *The power of geographical thinking*, Dordrecht: Springer, 1–6.
Dewey, J. (1990) *The school and society/the child and the curriculum*. (originally published 1900/1902). Chicago, IL: The University of Chicago Press.
Dewey, J. (2007) *Experience and education*. (originally published 1938). New York: Collier Books.
Edelson, D.C., Shavelson, R.J., Wertheim, J.A., Bednarz, S.W., Heffron, S. and Huynh, N.T. (2013) A road map for 21st century geography education: Executive summary. *National Geographic Society*. Dordrecht. https://mfpe.eventready.com/docs/download/Submission/Handouts/5358.pdf
GA. (2013) Thinking geographically. https://geognc.files.wordpress.com/2013/08/thinking_geographically.pdf
Harvey, S. and Daniels, H. (2009) *Comprehension and collaboration: Inquiry circles in action*. Portsmouth, NH: Heinemann.
Jackson, P. (2006) Thinking geographically. *Geography*, *91*(3), 199–204.
Karkdijk, J., van der Schee, J. and Admiraal, W. (2013) Effects of teaching with mysteries on students' geographical thinking skills. *International Research in Geographical and Environmental Education*, *22*(3), 183–190.
Kidman, G. (2018) School geography: What interests students, what interests teacher? *International Research in Geographical and Environmental Education*, *27*(4), 311–325.
Lambert, D., (2011) Reviewing the case for geography and the 'knowledge turn' in the English national curriculum. *Curriculum Journal*, *22*(2), 243–264.
Leat, D. ed., (2001) *Thinking through geography*. Cambridge: Chris Kington, 1998.
Mevarech, Z.R. and Kramarski, B. (2014) *Critical maths for innovative societies: The role of metacognitive pedagogies*. Paris: OECD.

Perry, J., Lundie, D. and Golder, G. (2018) Metacognition in schools: What does the literature suggest about the effectiveness of teaching metacognition in schools? *Educational Review*. DOI: 10.1080/00131911.2018.1441127.

Pickford, T., Garner, W. and Jackson, E. (2013a) *Primary humanities: Learning through enquiry*. London: SAGE Publications.

Reimers, F., Chopra, V., Chung, C.K., Higdon, J. and O'Donnell, E.B. (2016) *Empowering global citizens: A world course*. North Charleston, SC: CreateSpace Independent Publishing Platform.

Roberts, M. (2013) *Geography through enquiry: Approaches to teaching and learning in the secondary school*. Sheffield: Geographical Association.

Ruggiero, V.P. (2012) *Beyond feelings: A guide to critical thinking*. New York: McGraw-Hill Companies Inc.

Scoffham, S. (ed) (2010) *Primary geography handbook*. Sheffield: Geographical Association.

Smith, V.G. and Szymanski, A. (2013) Critical thinking: More than test scores. *International Journal of Educational Leadership Preparation*, 8(2), 15–24.

Van Laar, E., van Deursen, A.J., van Dijk, J.A. and de Haan, J. (2017) The relation between 21st-century skills and digital skills: A systematic literature review. *Computers in Human Behavior*, 72, 577–588.

Vygotsky, L.S. (1962) *Language and thought*. Ontario: Massachusetts Institute of Technology Press.

Vygotsky, L.S. (1978) *Mind in society: The development of higher psychological processes*. (translated by Cole, M.). Cambridge, MA: Harvard University Press.

Vygotsky, L.S. (1986) *Thought and language*. (translated by Hanfmann, E. and Vakar, G.). Cambridge, MA: MIT Press.

3 Teaching powerful geography through place

Introduction

The concept of place is at the core of geography. Places have a past, a present and a future. Children interact with places on a daily basis. They have strong feelings about places and they know what they like or dislike about any particular locality. Places are meaningful for children. A place-based perspective includes an analysis of children's emotional, aesthetic and spiritual connections to places and localities in addition to the manner in which their lives are shaped by the places in which they live. Place-based learning is defined as 'learning that is rooted in what is local – the unique history, environment, culture, economy, literature, and art of a particular place' (Rural School and Community Trust, 2005). While place-based learning has an obvious geographical focus, it is multidisciplinary and interdisciplinary.

This chapter sets out to:

- discuss what is meant by place, a sense of place and place-based education;
- demonstrate how to design resources including trail booklets for exploring local places;
- illustrate holistic frameworks such as the 8-way thinking approach for learning about places;
- examine place-based education though the case study of the Burren, County Clare, Ireland;
- explore the importance of place and place names in the context of identity;
- highlight the importance of learning to love our place;
- explore place-making initiatives such as story creation and curriculum making;
- highlight the importance of outdoor learning and a critical pedagogy of place; and
- showcase the potential of exploring local places with the help of well-known stories.

Place

Place is a complex phenomenon. Places can be local, immediate and concrete, such as a child's home, school and school grounds, or places can be far away and abstract. Buildings, towns, cities and countries are all referred to as places. Place can be as small as a room or as large as a continent. According to Rawling (2018: 55) 'place is an idea at the centre of arts and humanities scholarship, at the core of creative writing and innovation, fundamental to scientific understanding and at the heart of issues about identity, migration and conflict that threaten the stability and safety of our world'. Yet, Rawling contends that an understanding of place is given insufficient attention within geography. Nevertheless there have been some positive developments. Gruenewald (2003: 622) writes, 'place has recently become a focus for enquiry across a variety of disciplines, from architecture, ecology, geography and anthropology, to philosophy, sociology, literary theory, psychology, and cultural studies'. This is because, as the geographer Tim Cresswell (2015: 11) has written, place is 'a way of seeing, knowing and understanding the world'. One of the functions of primary geography is to develop children's fascination with their own and other places. Agnew (1987) describes three aspects of place as a meaningful concept: where a place is located, its setting (material or physical), and sense of place including the subjective and emotional attachment people have to and for places. These three aspects provide a useful framework for teachers as they plan for children to explore places.

Another threefold approach to studying place is illustrated in Table 3.1. Here I have taken a model developed for secondary geography (Cresswell, 2015; Rawling, 2018) and I have adapted this for primary geography. Through descriptive geography children learn about the unique features of a place. Social constructivist approaches explore processes

Table 3.1 Threefold approach to teaching place

Approaches to studying place (developed from Cresswell, 2015)	How place is understood (developed from Rawling, 2018)	Implications for primary geography
Understanding places (descriptive) Geographers aim to identify and describe each particular and unique place and to draw out its salient characteristics	Places are discrete areas of land with their own characteristics and ways of life (understanding of places)	Learning *about* place Emphasis on locational and place knowledge
Investigating processes (social constructionist) Geographers are interested in particular places in terms of underlying social processes, e.g. sustainability	Places are reflections of the processes and power relations that formed them (understanding of process)	Learning *in* place Place-based learning In the case of local areas this is a multidisciplinary way of children experiencing their local places In the case of other places this involves giving children an opportunity to vicariously experience place through: video clips guest speakers Skype calls images case studies
Exploring being in the world Geographers are interested in how place is an essential part of being human and how this is revealed, e.g. through creative responses to place	Place is a fundamental way of 'being in the world'	Learning *through* place Creating opportunities for children to experience place Local trails and field trips Use of artefacts Stories and literature Drama Music

Source: adapted from Cresswell (2015) and Rawling (2018)

impacting upon places. The act of being in the place recognises the strong relationship humans have with place. Rawling (2018) argues that school geography should include all three by providing opportunities for knowing and describing place, analysing and critically evaluating place and enjoying, sensing and being creative about place. For primary geography, I have taken this model and suggest that we should teach *about*, *in* and *through* place. Learning *about* place incorporates an emphasis on locational and place knowledge. Learning *in* place includes the philosophy of place-based learning discussed in detail later in this chapter. In the case of teaching about other places, this entails giving children an opportunity to vicariously experience place through: video clips, guest speakers, Skype calls, images and case studies. Learning *through* place involves opportunities for children to experience place through artefacts, local trails, stories, drama and music.

Sense of place

One of the core aims of primary geography is the development of a sense of place. Each of us has a relationship with the place we come from. These places shape who we are and how we see ourselves. The ability to be conscious of our place and to pay attention to its significance is referred to as having a sense of place (Feld and Basso, 1996). This process involves exploring – where places are, what they are like and how they relate to each other.

The term 'sense of place' is often used to emphasise that places are significant because they are the focus of personal feelings. Sense of place pervades everyday life and experience, and indicates that places are infused with meaning and feelings (Rose, 1995). A sense of place also incorporates spatial selection and environmental appreciation, as individuals are selective about the qualities which make their personal place special, situated and unique. According to Stedman (2002: 563), sense of place can be conceived of 'as a collection of symbolic meanings, attachments, and satisfaction with a spatial setting held by an individual or group'. Children's sense of place can be understood through an articulation of their 'special places'. Multi-modal presentations about favourite places include poetry, story, artefacts, images, short video clips recorded by children and interviews with local people. Children from Gaelscoil de hÍde, Oranmore, made presentations about their favourite places (Figure 3.1). The cumulative geographic learning by teachers and children was impressive.

Children have their own special places. They notice significant aspects and landmarks in their local area. As children explore their immediate environment, their innate sense of

Figure 3.1 Presentation about favourite places

curiosity is evident. Place attachment is important for emotional wellbeing. In Tanner's (2009) investigation of adults' special places, their responses related to one or more of three factors: fond memories of place, association with significant loved ones, and the intrinsic qualities of place, such as beauty. This illustrates that we have feelings about place and that some places are especially significant to us. These emotional bonds begin in childhood and are referred to as 'place attachment'.

Ask children to draw a picture or take a photograph of a place which is special to them. Then invite reasons for their selection. This can be extended through writing, poetry and/or art. Further extensions can occur through finding the place on maps of different scales, specifying map coordinates and using Google Street View for more images.

The local place is a phenomenal resource for primary geography. It provides an on-site laboratory for local investigations, promoting self-directed, enquiry-based transformative pedagogy. Children's local areas are part of their everyday geographies (Martin, 2008). These include:

- playgrounds, sports fields, parks;
- lakes, forests, woodlands;
- shopping centres, schools, street furniture, buildings.

Children have an immediate and special relationship with the place in which they live. It shapes who they are and how they see the world. From an environmental perspective, children can learn about what makes their place special, how to love and care for their place and how to become an ambassador for their place. From a more critical perspective, children can learn about the natural and human elements which shape their local places. The centrality of this concept is brought to the forefront of our education experience as we realise that we learn in and from places. As Greenwood (2013: 98) suggests, 'a theory of place that is concerned with the quality of human-world relationships first acknowledges that places themselves have something to say. Human beings in other words must learn to listen … '.

Children come from different places both within and beyond our country. It is a privilege to have a child from another country or culture in a class, not to mention the geographical potential for learning. The need for intercultural education in this era of globalisation has never been greater (Dolan, 2014). Helping children to understand and respect similarities and differences also helps the children to understand who they are in the context of race, ethnic group, culture, religion, language and history (Figure 3.2).

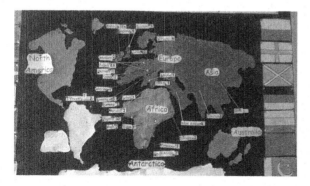

Figure 3.2 Holy Trinity NS, Mervue Galway, celebrates all cultures present in school

Case study 3.1 Trail booklets

Studying the local area is an important part of developing a sense of place. It is important to bring children outside for several different learning explorations. Street work, a distinctive form of fieldwork, focuses on urban areas where children explore streets and buildings. A trail is a structured walk through the local area with opportunities to observe and explore elements of the natural and built environment. Trails provide opportunities for children to observe, measure, record and describe human and physical features in their local area using a mixture of approaches including sketch maps, photography, writing, use of compasses and other equipment. While there is a strong geographical focus, trails provide opportunities to address other areas of the curriculum including history, literacy, maths and science. Scoffham (2017) provides invaluable advice on designing activities for urban areas such as exploration of street furniture, townscape features and conducting sustainable street surveys. Owens (2017) outlines opportunities for designing fieldwork based on children's questions and interests such as children working as location detectives. Local trails and field trips can become a springboard for extended investigation in terms of enquiry-based learning, cross-curricular projects and an analysis of local issues.

A trail booklet is a tailor-made or bespoke resource created specifically for exploring children's localities (Figure 3.3). Design of trail booklets is an example of curriculum making by teachers in consultation with children and the specific circumstances pertaining to the local area in question. Involving a number of teachers to design activities for local trails works well as a Continuing Professional Development (CPD) activity. Schools can also build up a bank of activities based on their local area appropriate for each age group or class level. Trail booklets also work best when the local area is viewed from a cross-curriculum perspective. There may be specific opportunities which lend themselves to mathematical, historical or scientific investigations. Trail booklets can be designed to facilitate 'explorations of the landscapes, soundscapes, and smellscapes', the school building, school ground and local area (Tuan, 2001: 5).

Ideally trail work should be conducted in three phases: preparatory work in the classroom, on-site work and follow-up activities and evaluation. A trail booklet can be designed to guide these three phases.

Main aims of trail work:

- To engage children in an exploration of their local place
- To help children observe, recognise and be critically aware of what is around them
- To increase awareness and understanding of the various human and physical features in their locality
- To arouse curiosity about local environmental issues
- To make appraisals about the condition of local areas
- To develop a range of skills such as observing, questioning, investigating and analysing
- To help children to experience their local area through different senses.

Prompts for planning a trail booklet

Character and nature of place

What do the children already know about this place?
What are their big questions about this place?

How would you describe the character or feel of this place? What is special about this place?
What words best describe this type of place, e.g. urban, rural, residential, village?
If you were describing/sketching this place for your local newspaper how would you describe this place?
Re-name this place with a name which captures the unique essence of the area.
What does this area look like and feel like?
Map the local area.
Assess the impact of weather on the local area.
How do we think and feel about our place?
What is important to people about our locality?
What sort of character does our locality have?

Noticing and recording changes in the locality

Where is the evidence of history in our place?
How has our area changed recently?
What was this place like in our parents' time/in our grandparents' time?
How may it change in the future?
Who makes decisions for our place?
What kind of changes would we like to make?
How can we communicate these changes?
What is changing in our place?
What can we see that shows how our area has developed?
What is the impact of changes?
How does this place change throughout the seasons?

Map work

Make a map of the area covered by the trail, marking significant landmarks.
Colour-code this map to illustrate how land is used.
How is land used around the school, near our homes and in this local area?
Use of the local area: design a list of activities in this area for children, teenagers, parents, retired people.

People in this place

Who lives in our area?
What do they do and like to do locally?
What are their opinions about this place?
What are their connections with other places?
What facilities and services are here?
Set up a jobs centre outlining the kinds of work available in this area. Invite parents to come into class to share experiences of their jobs. Bring children to employment sites: local hospital, fire station, post office, hotel, etc.
What do children/adults do for leisure? Map the leisure facilities in this area.

Taking care of our local area

How do we look after our local area?
In what ways have people affected our locality?
What changes would we like to see in the future and why?

Exploration of place through our senses

Sight: What are the noteworthy sights in this area?
Sound: Collect and record sounds.
Touch: Describe different elements in terms of the feelings associated with this place.
Collect smells.
Taste: Write a recipe with ingredients from this place.

Colours of place

Document the colour of our local place.
Design a palette of colours based on our local area.

Access

Identify connections between this place and other places in terms of routes, connections, roadways, connecting routes, etc.
Look at access for people with buggies, wheelchairs and walking aids.

Travel and transport

How do people travel?
Construct a travel survey.
How are vehicles powered?
Is there a petrol/diesel station in the local area or are there charging locations for electric cars?
Availability of public transport.
Document the availability and location of public transport facilities.

Improvement of place

Assess how the local area looks on the day of the walk.
Evidence of innovative practices which promote a sense of pride in local area, e.g. window boxes.
Evidence of litter and pollution.
What can be done to improve the area?
Conduct a litter clean-up.
Design improvement plans.
Class might sponsor a local sign, flower patch or information board.
What else can we tell a visitor about our place?
What would we show a visitor about our place?

Figure 3.3 Samples of trail booklets designed by student teachers

Place-based education

Place-based education (Sobel, 2004) involves designing education programmes to make school-based learning more relevant to everyday life through a focus on the local place and local issues. It uses place as the starting point for teaching language, arts and other subjects across the curriculum. It refers to education programmes whereby children learn about their local, natural, human, social and environmental area through enquiry, environmental action and other practical activities (Dolan, 2016b). A central concern is the establishment of a connection between children and their local area through experiential activities such as map making, field trips, conversations and 'hands-on real-world learning' (Sobel, 2004: 7).

The ultimate goal of Place-based Education (PBE) according to Israel (2012: 76) is to change the way children 'feel about and act in the places of their everyday lives in order to promote a more just and sustainable world'. PBE provides a framework for making connections between content and pedagogy, between geography and other disciplines, and between children and the world in which they live. Sobel (2004) cites many benefits of place-based education including increased academic achievement, the creation of stronger links with one's local community, an enhanced appreciation of the natural world and a greater inclination to working as an active citizen. Place-based learning makes a connection between cognitive learning, affective development and ethical actions. This according to Israel (2012) is essential for socially engaged geography educators.

While the term 'place-based education' is relatively new, the concept is rooted in Dewey's theories. Dewey (1940: 39) wrote that 'the school must represent present life as real and vital to the child as that which he carries on in the home, in the neighbourhood, or in the playground'. An educational framework that emphasises the role of teaching and learning in nearby places, hence connecting schools with the community and society, is place-based education. According to Hutson (2011: 19) 'place-based education seeks to connect learners to local environments through a variety of strategies that increase environmental awareness and connectedness to particular parts of the world'. McInerney et al. (2011: 5) suggest that the rationale for adopting place-based education in schools is that it 'creates opportunities for young people to learn about and care for ecological and social wellbeing of the communities they inhabit and the need to connect schools with communities as part of a concerted effort to improve student engagement and participation'. They also suggest that place-based education may position young people as producers rather than consumers of knowledge and provide them with knowledge and experience to participate in democratic processes.

Children have an immediate and special relationship with the place in which they live. It shapes who they are and how they see the world. From an environmental perspective, children can learn about what makes their place special, how to love and care for their place and how to become an ambassador for their place. From a more critical perspective, children can learn about the natural and human elements which shape their local places. The centrality of this concept is brought to the forefront of our education experience as we realise that we learn in and from places.

Understanding place: taking an 8-way-thinking approach

Gilbert's 8-Way Thinking (2006) provides a framework for thinking based on Howard Gardner's multiple intelligence (MI) theory and Philosophy for Children (P4C). It challenges and supports learning by engaging the learner with their eight intelligences, with the help of a simple yet powerful thinking tool. By working systematically through each of Gardner's intelligences and asking questions, children and teachers are able to think differently about a place or a geographical topic. The language of Gardner's intelligences has been simplified by Gilbert (2006) into the following easy-to-understand terms: *People, Numbers, Words, Nature, Sounds, Feelings, Sights, Actions*, as illustrated in the list in Table 3.2 and as an octagon in Figure 3.4.

Table 3.2 Developing enquiry questions

Intelligence	Shorthand	Enquiry questions might be
Linguistic intelligence – words and language	Words	What is this place called? How can we describe it with words? What do local signs tell us about this place?
Logical-mathematical intelligence – logic and numbers	Numbers	How can we use numbers to describe and explain this place? What can we count, measure, calculate? What can numbers tell us? Is there any evidence of numbers in this place? What do they signify?
Spatial intelligence – images and space	Sights	What can we see? What images best represent this place and why? How is this place mapped? How could it be mapped?
		What are the most significant sights in this place? How would you describe one of these sights?
Bodily-kinaesthetic intelligence – body movement control	Actions	What can you do here? What are other people doing? What does the landscape allow you to do? How do humans interact with the local environment?
Musical intelligence – music, sound, rhythm	Sounds	What can we hear here? What do sounds tell us about this place?

Interpersonal intelligence – other people's feelings	People	What do other people think about this place? How much do they value it and why?
Intrapersonal intelligence – self-awareness	Feelings	How do I feel here?
Naturalist intelligence – natural environment	Nature	How is the natural environment being sustained or improved? What natural features are there?

Source: based on Gilbert's 8-Way Thinking framework (2006)

The geographical issue or place under investigation can be placed in the middle of the octagon. Each section provides a new lens through which a geographical topic might be explored.

Incorporating the work of Rawlinson et al. (2009) and the 8-Way Thinking approach, I have worked with teachers and student teachers in developing place-based lesson plans and schemes of work (Figure 3.5). The framework enables us to study an area from many perspectives, which contribute to a layered understanding and rich appreciation of place. The framework also enables us to develop a bank of unique, tailor-made resources for each teacher, class and geographical area

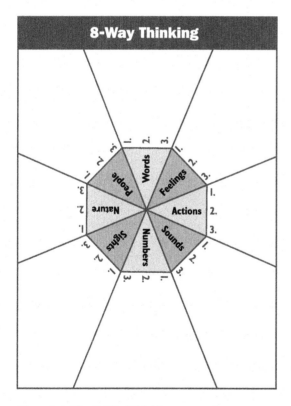

Figure 3.4 The 8-Way Thinking approach

Copyright owned by Ian Gilbert, CEO and Founder of Independent Thinking Ltd.

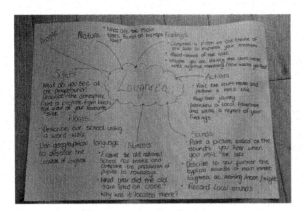

Figure 3.5 Use of 8-Way Thinking framework for planning geographical investigations in the locality

Case study 3.2 The Burren National Park as a case study in place-based learning

The Burren National Park

The Burren National Park is located in Co. Clare in the west of Ireland. The word 'Burren' comes from the Irish word *boíreann* meaning a rocky place. This is a very appropriate name considering the lack of soil cover and the extent of exposed limestone pavement. The Burren is world-renowned for its unusual flora, striking glacio-karst scenery and the density and quality of the archaeological heritage. It has many different fascinating and beautiful habitats, such as: species-rich limestone grasslands; hazel/ash woodlands and limestone pavements; turloughs; lakes; petrifying springs; cliffs and fern.

The Burren is at the heart of a network of learning communities that are exemplars of social transformation through learning. As a result, Burren communities are thriving and the rich natural and cultural heritage of the Burren is appreciated and celebrated as never before. Place-based Education (PBE) is used to situate the Burren at the centre of learning for many local primary schools.

Burrenbeo Trust is a landscape charity dedicated to connecting all of us to our places and our role in caring for them. It is dedicated to the Burren and its people with a programme which includes education, research, advocacy and conservation. The team of educators run a number of education initiatives for children, teachers and local schools. This includes an annual Learning Landscape Symposium with the focus on connecting people to places through place-based learning. They work with five primary schools each year as part of the Áitbeo schools programme and with secondary schools as part of the Ecobeo programme. Its engagement with primary and secondary school children through Ecobeo and Áitbeo is a defining commitment in creating, from an early age, a strong sense of identification with and care for their own place. Orientation sessions are provided for visiting groups, children and adults at selected locations across the Burren. Teacher education courses for primary teachers based on building place-based learning are provided annually. Burren Wild Child school tours and summer camps are designed with the main focus on learning about the environment through games and activities.

In Mary Immaculate College, student teachers spend an extended amount of time in the Burren as part of their geography elective (see images in Section 1 of the colour plates). Inspired by Payne and Wattchow's (2008: 6) 'slow or eco-pedagogical' approach students are encouraged 'to pause or dwell in spaces for more than a fleeting moment', therefore 'enabling them to develop place attachments and make meaning within the landscapes they inhabit'. The following is a range of comments from student teachers as they reflected on their experience:

> Over all I feel as though the trip to the Burren was extremely beneficial to my teaching of geography in the future. I have seen the importance for children to get out and explore their local areas through first-hand experience by me actually having the opportunity to participate in such activities. The ideas and theories I have come across in geography lectures and various readings on place-based learning make perfect sense after the trip. Children learn about their local place through experience and all of the activities during the trip to the Burren act as a tool to enhancing a child's sense of place.

> In my opinion it was essential that we took part in the activities during the trip. By participating, as teachers we were enabled to see the full extent of the learning taking place while doing the activities. Another benefit of taking part in the activities rather than just being provided with the instructions for each activity and game is that I remember the activities and will therefore be more likely to carry them out when I am teaching. The mapping activity was very versatile and could be used when exploring any place, be it rural or in a city.

> By exploring the surrounding area I learned things I did not know before. This highlighted the importance for children to explore their surrounding areas and showed me how much they can actually learn from engaging with their local place.

> It was clear to me from reading the articles and participating in the trip to the Burren that one is more engaged and eager to learn when experiencing nature. I can see how the level of enjoyment and engagement increases when the place children are learning about is familiar to them, children recognise certain aspects of the place, they are able to build on their prior knowledge and are able to relate to the content of learning. The learning becomes more personal for the children and ultimately, more geographical knowledge and skills are developed. All of the activities in the Burren were heavily based on place-based learning and developing a sense of place. I have gained some very valuable ideas and methodologies from my experience which I undoubtedly plan on incorporating into my teaching in the future.

Place and identity

Place is intertwined with our identity. In an increasingly mobile and globalised country like Ireland, a sense of place is still a strong marker of identity and is central to people's knowledge and understanding of themselves and others. When we are asked to introduce ourselves, we state where we come from. Place attachment is central to self-identity, providing a sense of security and stability. According to Tuan (2001: 9) 'we mark events in the timeline of our lives according to places, thereby making those places a part of our identity'. Higgins (2009: 48) suggests that developing a connection with place 'provides a start point for relationships (connections) with people within a community that allows further developmental outcomes, such as understanding the consequences of one's actions and an ethic of citizenship and care'.

Table 3.3 Examples of Irish place names

Irish term	Meaning	Examples of place names in Ireland
Cluain	Meadow	Clones, Clonmel, Clontarf, Clontibret
Béal	Mouth (usually of a river)	Ballydehob, Ballyshannon
Baile	Townland	Ballinasloe
Inis	Island water meadow	Ennis, Lahinch, Inis Meáin
Mam	Mountain pass	Maum, Maumturk

Place names

The study of place names can make a valuable contribution to PBE and local geography (Dolan, 2016a). In Great Britain and Ireland the significance of landscape is recorded in place names (Table 3.3).

Landscape features in Irish place names

MacFarlane's book (2015) *Landmarks* is a celebration of the language of landscape. By enriching our vocabulary of land terms, the language enhances our perception of and attentiveness to place. He makes a passionate call for restoring the 'literacy of the land'. In his book, he celebrates 'word magic' terms that can 'enchant our relations with nature and place'. Such vocabulary has been central in navigation, working the land and creative endeavours such as poetry writing. *Landmarks* presents hundreds of words and phrases for weather, natural phenomena, physical features and for working and playing in the countryside.

MacFarlane was shocked when he discovered words such as *ash, acorn, bluebell, otter, kingfisher* and *heron* were removed from the 2007 edition of the *Oxford Junior Dictionary* because of their perceived irrelevance. New words such as *blog, broadband* and *chatrooms* were introduced. He argues that, while there is a place for both lexicons, our lives are unquestionably impoverished without words such as *conker, lark* and other words related to our natural world. *Landmarks* includes themed glossaries of hundreds of these rare, deeply local, poetical terms, organised by such geographical terrains as flatlands, uplands, waterlands, coastlands, woodlands and underlands. The last glossary, *Childish*, is left blank, inviting readers to construct their own response.

In previous research, I refer to children as curators of place names (Dolan, 2016a). Through enquiry-based learning children can research local place names, interview local people, and document related local features. This approach casts children in the role of knowledge makers and local experts (Figure 3.6).

Place making, story creation and curriculum making

Further exploration of this interdependence between place, space and the environment can be achieved through a range of place-making activities in schools. Witt (2017) calls for a playful approach to geography. She writes about place making in the outdoors in general and through den making in particular as a means of providing children with opportunities to engage physically and cognitively with a place. According to Witt (2017: 50) 'den making is a universal experience of childhood'. Children enjoy creating their own places

Figure 3.6 Place name research

outdoors and these places can provide rich contexts for learning (Figure 3.7). Catling and Pickering (2010) argue that place making for children actively develops their higher-order thinking skills. Den making helps children assess the quality of their local area and consider their role in their community.

The experience of place and the expression of place can also feature through a variety of visual and storytelling methodologies. By giving places a story, children can understand what has happened to a place (geographically, environmentally, culturally, historically) and can develop their own connection with a place. Witt (2017: 55) argues that children should have opportunities to 'construct and respond to their surroundings, enabling them to build on and extend their personal geographies and celebrate the wonder of the world whilst creating a lasting sense of stewardship for their world'.

Place making is a universal act. Home owners put their individual designs on their homes. The Tidy Towns organisation encourages people to take pride in their local areas. When students move into university accommodation they put up photos and personal artefacts. Countries communicate a sense of place through signs of national identity – through national monuments, sports events, national celebration days and postage stamps.

Figure 3.7 Den building

Learning to love my place

Place-based education is about generating a deep knowledge of a particular place so that children will eventually care about landscape, nature and the people linked to a place. Stone (2009: 13) claims that 'places known deeply are deeply loved, and well-loved places have the best chance to be protected and preserved, to be cherished and cared for by future generations'. When people acquire a deep knowledge of a particular place, they begin to care about what happens to the landscape, creatures and people in it. Tovey (2007) suggests that we must allow children the opportunity to develop an intimate relationship with nature to understand but more importantly to feel the interconnectedness of all living things and to see their own place in the world. This is even more challenging considering that research indicates that teachers are working with children who have experienced a denaturalised childhood (Louv, 2010). This disconnect begs the questions: how do we encourage children to love something they do not know and how do we redress the imbalance?

To begin, we need to start thinking about places. Adopting place-based curiosity involves asking questions and thinking about places and our relationship with place. However, 'for many teachers and children, place remains merely a convenient container for factual details about different parts of the world' (Rawling, 2011: 70). Beames et al. (2012: 41) suggest that 'perhaps the underlying assumption behind place-based education is that directly interacting with "place" will foster an appreciation of and a broad ethic of caring for the land and its varied inhabitants'. Sobel (1996: 10) concurs: 'what's important is that children have an opportunity to bond with the natural world, to learn to love it, before being asked to heal its wounds'. Learning to love our area implicitly involves creative enterprises such as storytelling, art and poetry. While geographical in its own right, learning to love our local place is a cross-curricular endeavour. In the words of Demarest (2015: 70) 'when the story of a place emerges, subjects can merge beyond recognition'.

In British and Irish literature the association between literature and landscape is evident. There is an intense sense of place in Irish literature. Poetry, for example, is grounded in place. In his lecture entitled 'The Sense of Place' Seamus Heaney argued for the importance of place in the Irish oral and literary traditions. He suggested that 'we have to understand that this nourishment which springs from knowing and belonging to a certain place and a certain mode of life is not just an Irish obsession' (Heaney, 2014: 136). Children can explore the portrayal of landscape features in literature and they can create their own responses to local landscape through writing and poetry (Figure 3.8).

Figure 3.8 Children's place-based poetry

Baba Dioum (1968, cited in Norse, 1993: 193), a Senegalese naturalist and poet, highlights the need to focus on love and attachment in environmental endeavours when he states:

> In the end, we will conserve only what we love.
> We will love only what we understand.
> We will understand only what we are taught.

Outdoor learning

For children, learning to love places begins by bringing them outside. Outdoor learning provides children with an opportunity to experience the interdisciplinary nature of the real world through interactions with each other and the planet. Paradoxically, the literature suggests that children are spending less time outdoors (Gill, 2007). Curriculum theory for primary education has been influenced by the child-centred approaches of Steiner, Froebel and Montessori (Bruce, 2012). Froebel pioneered play as learning and outdoor play was a central element in his philosophy of learning. Montessori developed ideas about outdoor sensory play and the importance of using natural materials for teaching children. Steiner education incorporates outdoor time both for informal play and for the formal learning curriculum. These philosophical theories highlight the importance of outdoor learning for children in all curricular areas including geography.

The outdoor environment is a substantial but often underappreciated resource for primary geography. There are indications that opportunities for geographical fieldwork are diminishing (Ofsted, 2011). Many factors work in complex ways in locally situated contexts to influence when, if and how teachers conduct work outside the classroom. Financial constraints, time, teacher–pupil ratios, perceptions about safety, weather, transport, disruption to classes and teacher qualification all have an impact (Higgins et al., 2006; Mannion et al., 2013). Parental concerns about children's safety are reducing opportunities for children to independently explore their local area (O'Keeffe and O'Beirne, 2015). Children's play is increasingly moving indoors and onto electronic screens (Sobel, 2001). Concerns about children's current and future relationships with the environment are well documented. Louv's (2010: 36) discussions about 'nature-deficit disorder' describe the human cost of alienation from nature, including physical and emotional illness, reduced use of the senses and attention difficulties.

Despite the well-known benefits of outdoor learning, student teachers remain somewhat nervous about bringing children outside. The following letter from a second-year student teacher describes the benefits experienced from bringing children outside.

> Dear Anne,
> I just wanted to fill you in on an amazing experience I had today teaching geography while on school placement. I have to say I was skeptical when you introduced us to the concept of journey sticks in geography back in semester 3, but I definitely proved myself wrong with regards outdoor learning!
>
> Yesterday's weather inspired me to bring the children outdoors today during geography and I went about changing my lesson plan in order to do so. It was incredible! It was the most fun and valuable lesson I have ever taught, and neither of the other teachers had to step in once, although it was a great comfort to have them there. All I did was take the children out to the school garden, but even this had amazing

learning potential. I did an entire lesson on soil and its different layers, how water/fertilizer changes the texture of soil and also soil erosion (which then linked back to my theme of the Amazon Rainforest!). Not only did the children learn so much without lifting a pen but I also created such a good bond with them, because I was doing the work with them in the garden. I cannot explain how much this impacted upon my teaching!

Please keep encouraging student teachers to go outside with their classes. The fresh air alone does them so much good. Out of all of my classes, my current one is probably the one that has the most behavioural and intellectual needs, and therefore the one that I least expected to be taking outside. However, having seen the benefits I now think it is the most important way to reach the children's level and really create a connection between teacher and child that will aid their learning.

Thanks again,

This letters highlights the benefits of outdoor learning together with teacher anxieties about bringing children outside. This anxiety cannot be underestimated. It is a real factor present in many classrooms, keeping children firmly inside away from the messiness and the perceived unsafe nature of the real world.

Critical pedagogy of place

Building connections between children and others and between children and the places in which they live is at the heart of place-based learning. Place-based education is seen as an intervention in the world beyond the classroom. Gruenewald (2003) takes this further saying that children should do more than learn about places. He calls for a critical pedagogy of place which promotes social and ecological justice. Israel (2012) argues that place-based learning could make geography and geographical field work more powerful.

Place-based education, however, can be considered a challenge for educators. Place is an ideological construct. bell hooks (2009) demonstrates how the politics of race, gender and class affect the degree of place attachment experienced by groups and individuals. Many urban and rural places are far from idyllic; therefore, it is important not to over-romanticise the concept of place (McInerney et al., 2011) However, I believe that a geographical perspective can help children to consider what needs to be changed in their local area, thus giving children a sense of agency and voice. Another danger may be the temptation to accept the status quo and leave community prejudices and unequal practices unchallenged. When children and teachers view their schools and local areas through a critical lens they might ask questions such as those suggested by McInerney et al. (2011):

- What are the human and physical features of our local area? What can be done to improve our area?
- What do monuments, public architecture and signs tell us about heritage in our area? Whose voices are not represented?
- What could we do to ensure a more accurate record of past and present issues in our school and community?
- To what extent does our school model good environmental practices?

Teaching powerful geography through place 75

- What are the social, economic and cultural assets of our community? How fairly are they distributed? What can we do to work for a more just community?
- Who gets to make the decisions in our community? Are we as children consulted? What might we do to achieve a more democratic community and country?

A critical approach to PBE must focus on what needs to be protected in our community as well as what needs to be transformed. Focusing too strongly on the negative aspects of our local area can be counter-productive as this may ignite feelings of hopelessness. A sense of agency can be developed when children work as 'researchers, meaning makers and problem solvers' as part of a transformative approach to education (Demarest, 2015: 1). Place-based education begins with outdoor explorations as illustrated in Case study 3.3. and Figure 3.9.

Case study 3.3 We're going on a bear hunt!

In school young children begin their geographical explorations by becoming familiar with their school and school grounds. Senior infants (5–6 years) from Holy Trinity NS, Mervue, Galway explored their school and school grounds by conducting a bear hunt. Following paw marks and clues their adventure brought them through the school, around the school building and through the playground until they found their bear (Figure 3.9). Along the way they used key vocabulary including the names of different parts of the

Figure 3.9 Going on a bear hunt

school building, e.g. *playground*, *library*, *secretary's office* and locational words, e.g. *beside*, *near*, *far*, *away*, *from* and *through*.

Their adventure was based on the picture book *We're Going on a Bear Hunt* (by Michael Rosen and Helen Oxenbury). Loved by every child, this book includes rhythm through repetition, beautiful pictures, musical sounds and a journey based on a bear hunt. *We're Going on a Bear Hunt* tells the story of five children and a dog who set out one morning on an adventure to catch a bear. They are determined ('We're going on a bear hunt'), confident ('We're going to catch a big one'), optimistic ('What a beautiful day!') and fearless ('We're not scared').

On the way, they encounter all sorts of obstacles they 'can't go over and can't go under' but just have to 'go through'. These obstacles include a field of long wavy grass (swishy swashy), a deep cold river (splash splosh), a field of thick oozy mud (squelch squerch), a big dark forest (stumble trip!), a swirling whirling snowstorm (Hoooo woooo!) and finally, a narrow gloomy cave (tiptoe! tiptoe!).

Suggestions for extended geographical and place-based exploration

Geographical motions

Use your voice expressively to re-tell the story. YouTube versions of the picture book are also available but are not as effective as a dramatic, enthusiastic teacher. Ask the children to suggest actions for each of the phrases or words which are repeated in the chorus, such as those suggested in Table 3.4.

Drama and role play

When the children are familiar with the chant encourage them to perform it, acting out all the different parts of the journey as they say the words and move with expression. Using material from the local environment, ask the children to create sound effects for:

Swishy Grass, e.g. run rushes or a broom across the floor;
Splashy River, e.g. shake water bottles;
Squelchy Mud, e.g. shake jelly in a bottle;

Table 3.4 Suggested actions to accompany the story *We're Going on a Bear Hunt*

What a beautiful day	Draw a big sun in the air with both hands
We're not scared	Hands on hips, shaking heads
Under, over	Use one hand to indicate the matching movement; both hands move as if separating a path through
Swishy swashy!	Move hands alternately, left right, as if making a path through the grass, or wave both hands and body from side to side
Splash splosh!	Use alternate hands (and feet) to mime stepping
Squelch squerch!	Make large, slow, high steps as if your feet are sticking in the mud
Stumble trip!	Step carefully, then quickly – make little jumps, twists and wobbles – sideways, backwards and forward
Hoooo woooo!	Bend forward and walk very slowly, as if the wind is blowing. Rub your arms with your hands to keep warm!

Tripping in the Forest, e.g. bang pot with a stick or a wooden spoon;
Howling Snowstorm, e.g. blow into the top of an empty bottle;
Tiptoe in the Cave, e.g. bang a pot softly with a stick or wooden spoon.

Locational words and sites

Focus on key words such as:

under
over
through.

The children can practise these actions with various obstacles, make drawings based on these words and compile a map based on symbols for *under*, *over* and *through*. Alternative scenarios can be created based on words: *beside, around, in, out* and *behind*.

Focus on the places explored by the family in the book including a field of grass, a muddy plain, a forest and a dark cave. What kind of places can we explore in this local area? Urban sites will include words such as *road, street, shop, building* and *post office*. Rural sites will include words such as *field, wall, path, bridge, crossroads, tree, wood* and *hedge*.

Mapping and landmarks

Rewrite the story using landmarks from your school and school grounds. Map this journey for a bear hunt in the grounds of your school. Children can discuss the landmarks they pass on their way to school, e.g. post office, Una Murphy's house and the local Spar shop. Draw a picture of each of these landmarks on small cards. Discuss and sequence the cards on a map. This represents a map of each child's journey from home to school. This can be extended by writing a poem/story using the map and adding adjectives to describe each geographical landmark.

Organise a bear hunt

Organise a bear hunt in your classroom/school. Hide some soft-toy teddy bears around the playground. Encourage the children to work as a team to find them.

Obstacle course

Set up an indoor or outdoor obstacle course based on scenes from the book. Ask the children what obstacles they are likely to encounter in their local area, e.g. crossing a road, jumping over a wall or walking through a gateway. Children can suggest ideas for alternative scenarios which can be added to the obstacle course.

Artistic mapping

Use the descriptive language as inspiration for painting the contrasting scenes from the story.

Build a habitat for the bear

Once the bear has been located the children can locate a suitable resting place on the school grounds.

Creating geographical stories

The children can create alternative scenarios based on words: *beside, around, in, out, behind*, etc.

Reinforcement of geographical language through art

Create a wall mural/display to depict the different scenes in the story:

- Long wavy grass;
- Deep cold river;
- Thick oozy mud;
- Big dark forest;
- Swirling whirling snowstorm;
- Narrow gloomy cave.

Create a sensory tray for the play corner in the classroom:

Green crepe paper/construction paper/real grass for grass;
Blue and aqua waterbeads or marbles for water;
Miniature pine trees set in playdough: forest;
Potting soil or brown playdough: mud;
Cotton wool: snow;
Brown construction paper/cardboard: cave.

Playdough can be used for keeping everything in place on the tray. A little bed can be added beside the sensory tray to give children an opportunity to complete the story.

Children are born explorers. They begin to develop their sense of place during early childhood through their five senses (Gandy, 2007). As infants and toddlers they watch, listen, feel, touch, smell and taste as they get to know the world around them. After their first day in school, they continue exploring, experimenting and making discoveries seeking information that builds upon what they already know. Memories created during these outdoor adventures last a life time. When children are provided with rich interesting experiences during their early years, they view learning as a positive experience and begin to develop skills and competencies which lay the foundation for lifelong learning. They are giddy with excitement when they embark on an adventure such as a bear hunt or a Gruffalo hunt in their local environment. Place-based education can be fun, cross-curricular and story focused.

Case study 3.4 Geographical adventures with Flat Stanley

Flat Stanley is an American children's book series written by different authors. The original book by Jeff Brown tells the story of Stanley Lambchop who is flattened after an accident. He survives and makes the best of his new 2-dimensional shape, entering locked rooms under the door, flying as a kite and travelling to California in an envelope. The Flat Stanley

Project is based on children sharing ideas and information about places using a carboard cut-out of the character. Children create place-based projects using their Flat Stanley (or Flat Sarah, or another female equivalent). This usually includes photos of Flat Stanley in different places, journal entries, PowerPoint presentations or a scrapbook compilation of Flat Stanley adventures. These projects illustrate the child's home place and aspects which are special about their place, for example playgrounds, school building, home and favourite places. Once the children have completed their place study, Flat Stanley is ready to visit another part of the country or indeed another country armed with lots of geographical data.

Children send their Flat Stanley to a contact in another place. In addition to an introductory letter about Flat Stanley, children ask the recipient to continue Flat Stanley's adventure by documenting his experiences in his new country. He can be returned to the sender with photos, souvenirs and some information about the host country or area. Some children participate in the Flat Stanley initiative in response to an invitation from a friend or relative. One boy from Gael Scoil de hÍde brought his paper friend along on every geographical outing (Figure 3.10).

Many schools including Feakle National School, County Clare have adopted versions of the Flat Stanley project into their digital, literacy and geography programmes as shown in Figure 3.11.

Figure 3.10

Figure 3.11 Examples of Flat Stanley projects

Conclusion

Place, as one of the most important interdisciplinary concepts of the 21st century should be at the heart of geography (Rawling, 2018). In light of Rawling's concerns about not developing an understanding of place, I suggest that educators should teach *about*, *in* and *through* place. This chapter provides a number of frameworks for place to be explored geographically not only through geography but in conjunction with other curricular frameworks as well. Place-based learning, trail booklets and the 8-Way Thinking framework provide opportunities for wider exploration of and deeper engagement with place. Field trips such as the visit to the Burren discussed in this chapter and outdoor activities such as a bear hunt give children an opportunity to experience a place through all of their senses. Critically engaging with a place encourages children to evaluate their place in terms of aesthetics, function, sustainability, strengths, weaknesses and opportunities for improvement. Geography has a very important role in teaching children to love their place and this can only happen when children experience their place, critique it and respond to it through a range of curricular frameworks.

Exercise 3.1 Personal reflection for teachers

To what extent does your school use your local area as a key resource for teaching primary geography?

What resources are available in your local community for engaging children in place-based learning?

What talents/strengths are available among staff members for promoting place-based learning?

What place-based activities could be conducted for each class level?

Using some of the ideas in this chapter design a short-term, medium-term and long-term plan for using your local area as part of your primary geography programme.

Further resources

Demarest, A.B. (2015) *Place-based curriculum design: Exceeding standards through local investigations.* London: Routledge.
Pickering, S. ed., (2017) *Teaching outdoors creatively.* London: Routledge.
Waite, S. ed., (2017) *Children learning outside the classroom: From birth to eleven.* London: Sage.

Websites

Council for Learning Outside the Classroom: http://www.lotc.org.uk/; Ireland Leave No Trace: www.leavenotraceireland.org/; Scottish Natural Heritage (SNH): www.snh.gov.uk.

Articles published in the Geographical Association's journal *Primary Geography*

Bowles, R. (2005) Picturing places. *Primary Geography, 58,* 28–29.
Dolan, A. (2016) Naming our places. *Primary Geography, 89,* 22–23.
Jeffries, K. and Rogers, J. (2005) Sensing places. *Primary Geography, 56,* 34–35.
Nelson, C. (2005) Place, bias and personal geographies. *Primary Geography, 68,* 20–21.
Penhallow, S. (2006) Place through literature. *Primary Geography, 59,* 12–15.
Swift, D. (2005) Valuing places, thinking geographically. *Primary Geography, 58,* 4–7.
Tanner, J. (2009) Special places: Place attachment and children's attachment. *Primary Geography, 68,* 5–8.
Underwood, L. (2005) Ranking places. *Primary Geography, 68,* 40–41.

References

Agnew, J. (1987) *The geographical mediation of state and society,* Boston, MA: Allen and Unwin.
Beames, S., Higgins, P., and Nicol, R. (2012) *Learning outside the classroom: Theory and guidelines for practice.* London: Routledge.
Bruce, T. ed., (2012) *Early childhood practice: Froebel today.* London: Sage Publications.
Catling, S. and Pickering, S. (2010) Mess, mess glorious mess. *Primary Geography, 73,* 16–17.
Cresswell, T. (2015) *Place: An introduction* (2nd ed.). Hoboken, United States: John Wiley & Sons.
Demarest, A.B. (2015) *Place-based curriculum design: Exceeding standards through local investigations.* London: Routledge.
Dewey, J. (1940) *Education today.* New York: GP Putnam's Sons.
Dolan, A.M. (2016a) Naming our places. *Primary Geography, 89,* 22–23.
Dolan, A.M. (2016b) Place-based curriculum making: Devising a synthesis between primary geography and outdoor learning. *Journal of Adventure Education and Outdoor Learning, 16*(1), 49–62.
Feld, S. and Basso, K. eds., (1996) *Senses of place.* Santa Fe, NM: School of American Research Press.
Gandy, S.K. (2007) Developmentally appropriate geography. *Social Studies and the Young Learner, 20*(2), 30.
Gilbert, I. (2006) *8 way thinking.* www.teachingexpertise.com (12).
Gill, T. (2007) *No fear: Growing up in a risk averse society.* London: Calouste Gulbenkian Foundation.
Greenwood, D. (2013) A critical theory of place-conscious education. In Stevenson, R.B., Brody, M., Dillon, J. and Wals, A.E. eds., *International handbook of research on environmental education,* New York: Routledge, 93–100.
Gruenewald, D.A. (2003) Foundations of place: A multidisciplinary framework for place-conscious education. *American Educational Research Journal, 40*(3), 619–654.

Heaney, S. (2014) *Preoccupations: Selected prose, 1968–1978*. London and Boston: Farrar, Straus and Giroux.

Higgins, P. (2009) Into the big wide world: Sustainable experiential education for the 21st century. *Journal of Experiential Education*, *32*(1), 44–60.

Higgins, P., Nicol, R. and Ross, H. (2006) *Teachers' approaches and attitudes to engaging with the natural heritage through the curriculum*. Perth: Scottish Natural Heritage.

hooks, b. (2009) *Belonging: A culture of place*. London: Routledge.

Hutson, G. (2011) Remembering the roots of place meanings for place-based outdoor education. *Pathways: The Ontario Journal of Outdoor Education*, *23*(3), 19–25.

Israel, A.L. (2012) Putting geography education into place: What geography educators can learn from place-based education, and vice versa. *Journal of Geography*, *111*(2), 76–81.

Louv, R. (2010) *Last child in the woods: Saving our children from nature-deficit disorder*. London: Atlantic Books, Limited.

MacFarlane, R. (2015) *Landmarks*. London: Penguin.

Mannion, G., Fenwick, A. and Lynch, J. (2013) Place-responsive pedagogy: Learning from teachers' experiences of excursions in nature. *Environmental Education Research*, *19*(6), 792–809.

Martin, F. (2008) Ethnogeography: Towards liberatory geography education. *Children's Geographies*, *6*(4), 437–450.

McInerney, P., Smyth, J. and Down, B. (2011) 'Coming to a place near you?' The politics and possibilities of a critical pedagogy of place-based education. *Asia-Pacific Journal of Teacher Education*, *39*(1), 3–16.

Norse, E.A. (1993) *Global marine biological diversity: A strategy for building conservation into decision making*. Washington, DC: Island Press.

Ofsted (2011) Geography – Learning to make a world of difference http://www.ofsted.gov.uk/resources/geography-learning-make-world-of-difference.

O'Keeffe, B. and O'Beirne, A. (2015) *Children's independent mobility on the island of Ireland study*. Limerick: Mary Immaculate College.

Owens, P. (2017) Creative fieldwork: Whose place is this anyway? In Pickering, S. ed., *Teaching outdoors creatively*. London: Routledge, 42–58.

Payne, P.G. and Wattchow, B. (2008) Slow pedagogy and placing education in post-traditional outdoor education. *Journal of Outdoor and Environmental Education*, *12*(1), 25.

Rawling, E. (2011) Reading and writing place: A role for geographical education in the 21st century? In Butt, G. ed., *Geography, education and the future*. London: Continuum, 65–83.

Rawling, E. (2018) Reflections on 'place'. *Teaching Geography*, *43*, 55–58.

Rawlinson, S., White, C. and Kotter, R. (2009) Living geography: 8ways fieldwork – Evolution & evaluation. Paper delivered at Geographical Association Annual Conference, 11–14 April 2009, Manchester.

Rose, G. (1995) Place and identity: A sense of place. In Massey, D. and Jess, P eds., *A place in the world?* Oxford: Oxford University Press, 87–132.

Rosen, M. and Oxenbury, H. (2014) *We're going on a bear hunt*. London: Walker Books.

Rural School and Community Trust (2005) Place-based education learning portfolio. Retrieved 19 May 2018, from www.ruraledu.org/rtportfolio/index.htm

Scoffham, S. (2017) Streetwork: Investigating streets and buildings in the local area. In Pickering, S. ed., *Teaching outdoors creatively*, London: Routledge, 26–41.

Sobel, D. (1996) *Beyond Ecophobia: Reclaiming the heart in nature education* (Vol. 1). Barrington, MA: Orion Society Great.

Sobel, D. (2001) *Children's special places: Exploring the role of forts, dens, and bush houses in middle childhood*. Detroit, MI: Wayne State University Press.

Sobel, D. (2004) *Place-based education: Connecting classroom and community*. Barrington, MA: Orion Society.

Stedman, R. (2002) Toward a social psychology of place: Predicting behavior from place-based cognition, attitude, and identity. *Environment and Behavior*, *34*(5), 561–581.

Stone, M.K. (2009) *Smart by nature: Schooling for sustainability*. Healdsburg, CA: Watershed Media.

Tovey, H. (2007) *Playing outdoors: Spaces and places, risks and challenge: CA*. Columbus, OH: McGraw-Hill International.

Tuan, Y.F. (2001) Life as a field trip. *Geographical Review*, *91*(1–2), 41–45.

Witt, S. (2017) Playful approaches to learning out of doors. In Scoffham, S. ed., *Teaching geography creatively*, Abingdon: Routledge, 44–57.

4 Playful approaches to powerful geography
Games, artefacts and fun

Introduction

Play-based learning has much to offer the teaching of geography throughout the primary years. Geographical play allows children to approach concepts and issues from different perspectives. Playfulness supports the development of creativity, an essential 21st-century skill. Its potential role in the geography classroom is significant. The inclusion of play helps to generate excitement, enjoyment and interest as part of the process of geographical learning. Playful learning can be defined as 'learning experiences that are child-led or adult-initiated or inspired when children engage in playful ways' (Moyles, 2010: 21). To play is to engage. When children play they pick up objects and ideas, turn them around and change them into something magical, innovative and inspiring. Play helps children dare to learn even when they are uncertain; it is about what will happen next. It creates an attitude of mind 'which is curious, investigative, risk taking and full of adventure' (Bruce, 2012: 41). It allows children to work as young geographers in a quest of exploration and discovery.

There are many different types of play including constructive play, role play, imaginative play, rough and tumble play, socio-dramatic play and fantasy play. Geography provides a valuable context for play-based education. While a quality play-based education is recommended by early-year practitioners, it also has much to offer older children. However, opportunities for play in formal education tend to decrease as children get older. As Woodyer (2012) argues play is fundamental to human experience across the life course.

This chapter sets out to:

- discuss playful approaches to learning geography;
- describe geographical playfulness with reference to different types of play;
- highlight the importance of tactile experiences of geography through handling artefacts and culture kits;
- examine the opportunities of playing with rocks.

Geographical playfulness

Playful approaches to primary geography offer children opportunities to engage with their world through problem-solving, decision-making and first-hand practical experiences (Morse and Witt, 2014). Play influences all areas of development; it offers children the opportunity to learn about the self, others and the physical environment (Catron and Allen, 2007). Play allows children to use their imagination while developing their communication and their fine and gross motor skills. Physical, cognitive and emotional capabilities are also developed through play. Early geographical experiences, such as exploring spaces and manipulating objects in the environment, help children begin to understand the world around them. Older children can continue to manipulate objects through working with artefacts, culture kits and games.

Playfulness is associated with fun, creativity, problem-solving and child agency. Ackermann (2014: 1) suggests that being playful in creative processes is a necessary aspect since 'coming at things obliquely – through suspension of disbelief (pretence), artful détournement (displacements), and playful exaggeration (looking at things from unusual angles) – allows [one] to break loose from the habitual'. Csikszentmihalyi (1997) suggests that when children are involved in play, particularly when they are 'carried away' with what they are doing, (referred to as 'flow'), deep learning is taking place.

Playful approaches in education are located within an enquiry-based or constructivist theory of learning involving both experience and reflection as part of the process. Playful learning involves children as knowledge creators. The benefits of play are numerous including the promotion of creativity, imagination and spontaneous learning. These creative skills require a variety of approaches to allow children to explore issues from a variety of perspectives. Playful methods in particular encourage and enable alternative views to be produced and explored. According to Denmark's first Professor of Play, Helle Marie Skovbjerg (2018), 'play is often a matter of getting something in your hand and then exploring the world through it'. Play is thus an accessible means of bringing the world directly to children. Through play, children can work as geographers, exploring, observing and analysing their immediate environment.

Playfulness in the outdoors

Children develop their relationship with nature through play (Kalvaitis and Monhardt, 2012). Research shows that empathy with, and love of, nature grows out of children's regular contact with the natural world (Louv, 2010, 2013). Explorations, self-initiated discovery and hands-on interactions are the best way to inspire children, to cultivate a sense of place and to generate a sense of wonder and awe (Chawla, 2016). However, children's time outdoors has been decreasing dramatically for many reasons. Through place-based learning approaches, as discussed in Chapter 3, geography can cater for children to critically and playfully engage with their local environment. Nature provides for discovery, creativity and problem-solving by providing opportunities to observe, notice, think, question, offer solutions and be reflective. Children can draw in the sand, make designs with twigs, build a den with branches or simply lie on the ground and look at the sky. Preparation for all weather conditions will ensure children are always ready for outdoor adventures.

Aistear early childhood education framework

Aistear (the Irish word for 'journey') is the early childhood curriculum framework for all children from birth to six years in Ireland (NCCA, 2013). As the name implies, education is a lifelong journey that can take many different routes. This framework describes types of learning (dispositions, values and attitudes, skills, knowledge and understanding) that are important for children in their early years, and offers ideas and suggestions as to how this learning might be nurtured. The framework uses four interconnected themes to describe the content of children's learning and development: *Wellbeing, Identity and Belonging,*

Communicating, and *Exploring and Thinking*. The geographical relevance of these themes ensures that geographical play is well placed to support this early childhood framework.

The Aistear curriculum, which draws heavily on the New Zealand Te Whariki curriculum, is thematically based, and is influenced by emerging sociology of childhood perspectives. Aistear is committed to providing dedicated time each day for children to engage in play. It recognises the different kinds of play described in Table 4.1.

Table 4.1 Different kinds of play

Creative play involves children exploring and using their bodies and materials to make and do things and to share their feelings, ideas and thoughts. They enjoy being creative by dancing, painting, playing with junk and recycled material, working with play dough and clay, and using their imaginations.

Play through games with rules allows children to appreciate the advantages and disadvantages of rules. Young children begin to play by their own flexible rules. In time they participate in more conventional games with external rules. Language is an important part of games with rules as children explain, question and negotiate the rules. Rules are often an important part of pretend play where children negotiate rules about what can and can't be done.

Language play involves children playing with sounds and words. It includes unrehearsed and spontaneous manipulation of these, often with rhythmic and repetitive elements. Children like playing with language – enjoying patterns, sounds and nonsense words. They also love jokes and funny stories.

Physical play involves children developing, practising and refining bodily movements and control. It includes whole body and limb movements, coordination and balance. These activities involve physical movements, for their own sake and enjoyment. Children gain control over their gross motor skills first before refining their fine motor skills.

Exploratory play involves children using physical skills and their senses to find out what things feel like and what can be done with them. Children explore their own bodies and then they explore the things in their environment.

Manipulative play involves practising and refining motor skills. This type of play enhances physical dexterity and hand–eye coordination. Over time children need to experience a range of different levels of manipulation if they are to refine their motor skills. This type of play includes manipulating objects and materials.

Constructive play involves building something using natural and manufactured materials. As children develop, this type of play becomes more complex and intricate.

Pretend, dramatic, make believe, role and fantasy play involve children using their imagination. These types of play include pretending with objects, actions and situations. As children grow, their imaginations and play become increasingly complex. Children use their developing language to move from thinking in the concrete to thinking in the abstract. They make up stories and scenarios. Children act out real events and they also take part in fantasy play about things that are not real, such as fairies or super heroes. Children try out roles, occupations and experiences in their pretend play. Early literacy and numeracy are clearly evident in this type of play, for example, children make lists and menus and pay for cinema tickets.

Small world play involves children using small-scale representations of real thing like animals, people, cars and trains sets as play props (Figure 4.1).

Socio-dramatic play involves children playing with other children and/or adults. It provides opportunities for children to make friends, to negotiate with others, and to develop their communication skills. By extending children's language skills, the process of story creation becomes more elaborate.

Source: NCCA, 2013: 54

Imaginative play

Imaginative play allows children an opportunity to experiment with different experiences of interest to them. By fostering cognitive and social development, children make sense of their world through play. Children can identify with the adult world through role play by developing social skills through interacting with other children. By rehearsing real-life scenarios children build confidence, self-esteem, and communication skills. Through learning how to negotiate, they can develop their imagination, emotional intelligence and language skills. Greene (1995: 5) talks about the importance of our social imagination: 'the capacity to invent visions of what should be and might be in our deficient society, on the streets where we live, in our schools'. Futures thinking, which is critically important for geographers, involves imagining different scenarios (Hicks, 2014).

Imagination opens the door to possibilities. It is where creativity, innovation and thinking outside the box begin. During imaginative play, children express themselves verbally and non-verbally, experiment with different roles, plan intentionally and non-intentionally, manipulate materials and experience alternative ways of being. From a geographical perspective, playful and imaginative learning can improve children's motivation and observation, develop children's geographical thinking and promote enquiry-based learning (Witt, 2017). Children from St John the Apostle National School, Knocknacarra, Galway designed a 3-D model house and described its unique features in oral language. This work formed the basis of preparing a document for selling their wonderful house (Figure 4.1).

Figure 4.1 A model house designed by 6-year-old children in St John the Apostle NS, Knocknacarra, was subsequently advertised for sale by a local auctioneer

Figure 4.2 Playing with Rory's Story Cubes

Storytelling is one of the oldest forms of sharing thoughts, ideas and information. Children have the opportunity to create and co-create stories during playtime. Children can travel vicariously to new lands, experience fresh dilemmas and problem-solve their way through scenarios. Stories are set in places – real places, imaginary places, hidden places, secret places and virtual places. Each story takes place in a specific geographical setting, be that a farmyard, a forest or a beach. From a spatial or place perspective, children should be actively encouraged to name and describe the physical characteristics of their geographical setting.

When children are developing their stories some carefully constructed questions will help them develop a sense of place, for example:

What is this place like?
What makes this place special?
Tell me about this place.
What activities do people do here?
Is it hot or cold here?

Using the imagination to create geographical stories and language is greatly assisted by resources such as *Rory's Story Cubes*, as indicated in Figure 4.2. The game consists of nine dice, each one illustrated with a different image on every face. By throwing the dice and using prompts from the visual images children can create stories. One of the sets of dice (Voyager) has specific geographical images. The cubes are based on the idea that the brain thinks in pictures but communicates in words. Using images to trigger stories helps the brain think in new ways.

Playing with unstructured materials

Children naturally engage with geographical concepts when they create new environments informed by their imagination. Giving them an opportunity to discuss their creations and the dilemmas incurred maximises geographical learning. As explorers, children can examine new landscapes of forests, deserts and marine environments, thus building their understanding of the world. Significant opportunities for geographical learning are possible when children participate in creative play with dolls, vehicles, blocks, rocks,

Table 4.2 Examples of unstructured materials

Balls	Chalk	Jam jars and lids	Paper plates	Rope	Trays
Bark	Clothes pegs	Keys	Pasta	Sand	Twigs
Baskets	Clothes for dressing up	Leaves	Pebbles	Saucepans	Wallpaper
Blankets			Pegs	Shells	Water
Bottle caps	Cushions	Lentils	Pine cones	Spools and reels	Wooden utensils
Boxes (various sizes)	Egg cartons	Logs	Pipes and pipe cleaners	Sticky-tape	Wool
	Fabric	Magazines			
Bricks	Feathers	Newspapers	Plastic containers	Stones	
Buckets	Food containers	Nuts and bolts	Ribbons	Straws	
Buttons		Paper		String	

cardboard or boxes. Witt (2017: 47) argues that 'in order for playful learning to thrive, the environment needs to provide opportunities for children to manipulate, construct, observe, listen and touch their world'. Hence the classroom environment needs to be organised accordingly.

Creative thinking occurs when children manipulate play dough, create recipes by mixing dirt and water, work with art material and splash in puddles. Critical thinking transpires when children imagine and try new ways of doing things. Problem-solving skills are acquired as children experiment with different scenarios. Kress (2010) points out that the abstract aspects of teaching become tangible through different materials. Teachers foster geographical engagement through questioning, promotion of locational language and making connections between the children's experiences and geographical language.

Materials can help children to think geographically as they 'evoke memories, narrate stories, invite actions, and communicate ideas' (Pacini-Ketchabaw, et al., 2017: 1). Using open-ended materials gives children the freedom to choose and create, developing confidence and competence. Access to sticky-tape, scissors, paint, paper and glue extends creative opportunities for children. A simple length of fabric can be the sea, a river, a roof, the sky, or part of a den. Open-ended materials such as those suggested in Table 4.2 are suitable for use by all children as they will manipulate and use them according to their stage of development. Play with natural materials (sand, stones, pebbles, shells and water) allows children to explore the journey of a river, flooding and other natural disasters. Through working with materials children can learn how to problem-solve, investigate, discuss, share ideas and negotiate. The notion of 'an encounter with materials' suggests that both children and materials are transformed through a process of thinking with materials (ibid.).

Small world play: Getting up close and personal

Small world play is a type of imaginative play where miniature versions of scenes are created for children to play with, using small props and toys (for example, a farmyard scene with small toy farm animals). It enables children to create and manage different environments through storytelling, use of their imagination and acting out scenarios. By allowing

Table 4.3 Small world play resources are needed for both the small world inhabitants and the small world landscape

The small world inhabitants	The small world landscape
• People • Animals (pets, farm animals, zoo animals, wild animals, sea creatures, mini beasts) • Dinosaurs • Vehicles (cars, trucks, diggers, trains, tracks, planes) • Fairies and other fantasy figures • Aliens	• Dolls' house and furniture • Space • Lunar landscape • Garage • Train station • Shops • Castles and forts • Farms • Woodland • Jungle • Beach • Under the sea • The Arctic or Antarctic

Source: Early Childhood Ireland, 2015

children to take control, small world play supports the development of locational, spatial and geographical language. It allows children to express the world they carry in their heads (Barnes, 2003) and through this expression they can share their knowledge and further develop their understanding of place.

To facilitate small world play, resources are needed for both the small world inhabitants and the small world landscape (Early Childhood Ireland, 2015), as shown in Table 4.3.

Small world play creates endless opportunities for children's geographical learning, helping children to expand their imagination and be creative. It enhances language and storytelling skills through acting out everyday scenarios. When creating a small world consider the interests of the children. A favourite story can be extended into the small world area. Geographical settings such as deserts, rainforests and villages can be designed, created and explored by children.

Once the theme has been decided, materials need to be collected. Elements which support sensory play are particularly popular (water, sand, play dough, straw, uncooked rice, rocks and pebbles). With a variety of resources, children imagine, create and change different scenarios. They make roads, pathways, enclosures, green spaces and small villages as part of their small world site. Children can find solutions to problems such as locating the best site for a tent on a toy camping site. By asking the children to solve problems, small world play helps children describe what they see, suggest alternatives and see possibilities. Small world play with toy villages, dolls' houses and farm sets visually introduces children to concepts of scale (Figure 4.3). Children can draw picture maps based on their small world and compare their drawings with other children. Small characters and vehicles moved around by the children encourage the use of positional language.

Use of miniature figures in indoor and outdoor environments has been used extensively by Witt (2017: 64), who describes miniaturisation as 'a powerful pedagogic technique'. Her work has been inspired by the book *Little People in the Cities: The Street Art of Slinkachu* (2008). Miniaturisation enables children to look at a place from varied perspectives. Miniature figures can include Lego and Sylvanian Family characters. Journeys

Playful approaches to powerful geography 91

Figure 4.3 Small world play in fairyland

with miniature figures can be recorded by drawings, photographs or video and recreated using book creator apps.

Mission Explore

Sometimes the value of imaginative play is forgotten by teachers and student teachers. Teacher education is well placed to include geographical approaches whereby student teachers engage in activities so that they might 'look beyond things as they are' and 'anticipate what might be seen through a new perspective or through another's eyes' (Greene, 1995: 49). Mission Explore (Parkinson, 2009) is a new way of thinking and interacting with the world encouraging children to explore playfully and learn outdoors. Also known as guerrilla geography, explorers think of questions, set challenges, search for answers, make exciting discoveries and try to improve places. Hundreds of illustrated quirky missions have been published, challenging young people to rediscover their world. There are numerous opportunities to get to know local areas better and have lots of fun at the same time. Challenges include *Let a dog take you for a walk* and *Draw a local fantasy map*. Other samples are included in Figures 4.4 and 4.5. Mission Explore is about encouraging people to engage with geography creatively, rediscovering a sense of wonder while having as much fun as possible. The Mission Explore approach 'is embedded in exploring, questioning, playing, experimenting and experiencing our planet, while developing critical thinking around the interconnected, multiple and dynamic geographies that every aspect of our lives entails' (Askins and Raven-Ellison, 2012: 163).

Each year my students explore the campus using the Mission Explore initiative (Figure 4.6).

Figure 4.4 Sample from Mission Explore: Establish a new country
Copyright: The Geography Collective/Tom Morgan Jones 2010

Figure 4.5 Sample from Mission Explore: Go photo orienteering
Copyright: The Geography Collective/Tom Morgan Jones 2010

Figure 4.6 Images of student teachers on a mission to explore their campus geographically

Playing geographical games

There are several games available for promoting geographical thinking and learning (Whyte, 2017). Games allow children to learn through collaborative experiences, enhanced cooperation and teamwork (Tidmarsh, 2009). Children and teachers can design their own geographical games including: versions of Snakes and Ladders; Snap (with pictures and text describing geographical features); Who Wants to be a Millionaire?; and Monopoly (discussed below). Games allow children to develop problem-solving, oral language and thinking skills. Figure 4.7 provides an example of a game designed by student teachers.

Several board games have so much potential to make learning geography fun and engaging. Monopoly, for example, is one of the bestselling board games in the world. It can be found in no fewer than 114 countries across the globe and is played in 47 languages. While many of us have fond memories of playing the London version, trying to acquire property in Mayfair, there are now several versions of the game available in the UK, the USA and internationally. In Ireland, there are versions based on Galway, Limerick, Dublin, Cork and Belfast. In the Galway City version, the Latin Quarter, Galway United, Galway City Museum, the Galway Races, the Aran Islands and Galway's railway station

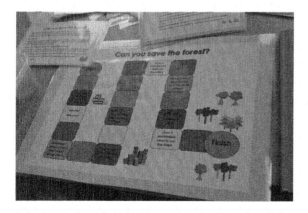

Figure 4.7 A game developed by student teachers

94 *Playful approaches to powerful geography*

and harbour are among other landmarks (and events) featured on the board. Geographical versions of the property board game have been created in other jurisdictions, including 'Myplace-opoly' (Woolliscroft and Widdowson, 2010) and 'Exe-opoly – an investigation of your city' (based on the city of Exeter) (Clemens et al., 2013). Earth-opoly is the Planet Earth edition. Rare photographs of the Earth's landscapes and animals decorate the game board. Players can travel and experience the Earth's natural habitats and elusive animals as they dive down the Cave of Swallows, migrate though the Gobi Desert, climb the Himalayas and encounter caribou, crocodiles, foxes, giant pandas, orangutans, seals and dolphins. Packaged in recyclable material, this special edition includes a 'Can you find it?' geography game and six custom pewter tokens featuring animals: elephant, kangaroo, penguin, polar bear, blue whale and amur leopard.

Monopoly has a square board. Each side has ten squares on which a player's piece can land. Each player has a token to mark his or her position on the board. The original game included ten metal tokens including an iron, purse, lantern, race car, thimble, shoe, top hat, battleship, cannon and rocking horse. Many more have been added and taken away over the last eight decades. A player rolls a pair of dice and moves the total of the two dice. Randomness is introduced to the game through two shuffled piles of card called *Chance* and *Community Chest*.

Playing the game involves making a variety of decisions such as buying or not buying property, houses and hotels, dealing with the banker and making trades. The point of the game is to acquire property, build houses and hotels, avoid jail and pass *Go* as many times as possible to collect extra money.

Making games is not a new idea in education (Walford, 2007). However, according to Kafai (2006: 36), 'far fewer people have sought to turn the tables: by making games for learning instead of playing games for learning'. Children can make their own Monopoly set based on their local area. They can make their own personalised version by using the traditional template of a Monopoly board (Figure 4.8). The game can be based on their school, local area, county or country depending on the geographical focus being pursued. In the case of the children's local area, they can design the template with the names of local streets, buildings, roads and landmarks. Real estate cards can be designed to match the names on the template. Children can use counters, dice, Community Chest cards, Chance cards, money, houses and hotels from a regular Monopoly game or they can make their own versions. Sixteen cards are required for both Chance and Community Chest

Figure 4.8 A geography game based on the Monopoly model

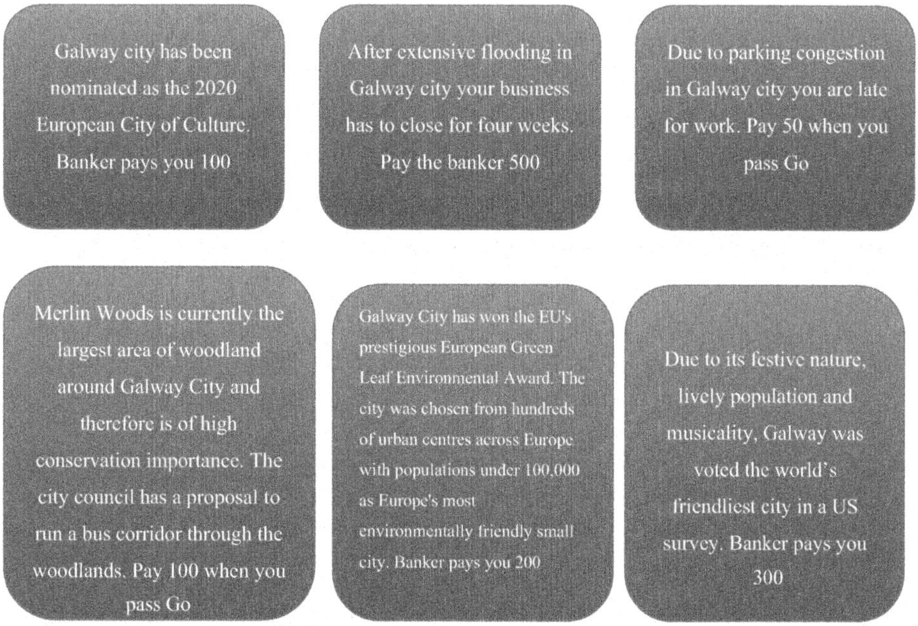

Figure 4.9 Examples of Chance cards for Gal-opoly game (version of Monopoly based on Galway city)

[Card 1: Galway city has been nominated as the 2020 European City of Culture. Banker pays you 100]

[Card 2: After extensive flooding in Galway city your business has to close for four weeks. Pay the banker 500]

[Card 3: Due to parking congestion in Galway city you are late for work. Pay 50 when you pass Go]

[Card 4: Merlin Woods is currently the largest area of woodland around Galway City and therefore is of high conservation importance. The city council has a proposal to run a bus corridor through the woodlands. Pay 100 when you pass Go]

[Card 5: Galway City has won the EU's prestigious European Green Leaf Environmental Award. The city was chosen from hundreds of urban centres across Europe with populations under 100,000 as Europe's most environmentally friendly small city. Banker pays you 200]

[Card 6: Due to its festive nature, lively population and musicality, Galway was voted the world's friendliest city in a US survey. Banker pays you 300]

cards. Sample card statements are set out in Figure 4.9. Card actions can be kept the same but customised for each local area. Cards can be designed for each set of property. For simplicity, use of the same rent and mortgage amounts as the corresponding cards from a regular Monopoly game is advised. The game board should include three Chance squares, three Community Chest squares and six Utilities (e.g. electricity provider, water, phone, railway, etc.) together with matching cards. The discussion involved in making an area-based Monopoly game involves local and place-based knowledge. The following discussion points make a good starting point:

What are the most popular areas in your local town/area?
What are the least popular areas in your town/area?
Make a list of 22 locations and rank them from the best to the worst places to live.
Put these locations on your Monopoly board (most popular places on dark blue section) and least popular places on brown sections.
Make Chance cards and Community Chest cards based on socio-economic factors or current issues in your local area (samples are provided in Figure 4.9).

Simulation games

Complex real-life issues such as trade, intercultural misunderstanding and unequal distribution of wealth can be taught through simulation games such as the World Trading Game (produced by Christian Aid). Through simulation games, children experience life-like situations in a realistic environment which is conducive to the active involvement of all.

Children can vicariously experience the rules and regulations of real-life events. Elements of conflict, competition and the prospect of winning scores encourage children to play. Simulation provides opportunities for active learning through cooperation, teamwork, analytical thinking, decision-making and problem-solving. They allow children to create and manipulate systems and problems within certain parameters and without the normal constraints of time and space. They are experiential, collaborative and learner centred. By helping children think outside the box they generate new insights and alternative perspectives of an issue and provide material for further exploration and analysis.

The World Trading Game is a simulation designed to introduce children to the realities of trade and how it affects the prosperity of a country. The game shows, through the production of different paper products, how the world trade system works, how countries interact, how poorer countries are more vulnerable to unfair practices and how richer countries are far more advantaged in terms of trade agreements. The game highlights how the gap between rich and poor nations is made wider by the trading policies of wealthier nations. It demonstrates how poorer countries, especially those selling raw materials, have great difficulty in obtaining a fair price for their goods. The game works well with children aged 11–14 years of age.

The Paper Bag Game (produced by Christian Aid) is an interactive game which gives children an insight into challenges faced by children trying to earn an income with few resources. The objectives of the Paper Bag Game are to help young people to explore some of the issues around child labour in India and to understand some of the pressures of trying to survive in an economy with massive unemployment and no social security. Using the real cost of living prices and wages, children are introduced to the idea of managing, if they had to survive, by making and selling paper bags. The game challenges children to examine our own use of the world's resources and how global economic systems exploit vulnerable communities.

Reaching out to the world through artefacts

We live in a 3-dimensional world. Being able to visualise and manipulate shapes and images mentally vastly improves one's spatial intelligence. The first step in this process involves physically manipulating and moving objects. The word 'artefact' derives from the Latin phrase *arte factum* meaning to 'skilfully make' or, in other words, an object created by human hand. Artefacts can be small (such as coins, receipts from a different town or country, postcards or images) or they can be larger (dolls, maps, games from other countries or a video clip).

Physical manipulation of objects, cards and materials can bring geography alive for children by making concepts accessible and meaningful and by removing the sometimes abstract, elusive nature of geographical knowledge. Handling objects generates curiosity, invites children to play and encourages enquiry-based learning. The main aim of using artefacts in geography lessons is to make places and geographical themes more accessible to children.

Artefacts can be used as a lesson starter, as a core resource for a series of lessons or for assessment. Geographical artefacts can lead to rich geographical discussions and research. They can also provide tactile experiences, which aids investigation, and they are particularly effective with younger children, children with special needs or children who struggle with literacy. A geographical artefact provides teachers and children with greater access to geographical knowledge through a process of observation, enquiry and investigation.

While artefacts are more commonly associated with history (Hoodless, 2008) and religious education (RE), they have a unique contribution to make in geography lessons.

Artefacts are rich sources of meaning which, through careful examination, can reveal information about place, social customs, arts, religious beliefs and culture. What often lies at the centre of a brilliant geography lesson is an artefact which is used purposefully. Whether to create the need to know or to illustrate geographical examples, artefacts can bring a lesson to life. Examples of geographical artefacts are set out in Table 4.4. The potential power of each artefact lies with the teacher. Creative uses of such resources sit at the heart of curriculum making.

Table 4.4 Types of geographical artefacts

Type of artefact	Geographical connection	Examples
Ephemera	Items of transitory use, usually discarded because they are disposable. They identify and inform children about the place in some way, for instance, its features, goods sold and events and activities seen, bought or undertaken	Local and national newspapers, timetables, shop till and transport receipts, sweet bar and other food wrappers, postcards, leaflets and cards, posters, plastic and paper bags, restaurant and takeaway menus
Purchased goods	Short- or long-lasting items, purchased for their usefulness or interest at the time or as mementoes of a particular place. They name or indicate what is available in the place or the wider local area	Clothing, tea towels, sun glasses, food ingredients and samples made locally or commissioned for sale in the area
Local publications	Publications which provide information for future reading about or as mementoes about the place and wider area visited	Tourist guides, locally and nationally available maps, books and booklets about the locality, walking guides, story and poetry books produced locally, parish and council or similar magazines, and local directories
Personal creations	Items that would not exist about the particular place or wider area had not the visitor created or taken them while on the visit	Personally taken or made photographs, sketches, drawings and sketch maps
Natural collection objects	Natural objects which it may be possible to collect from the environment (but often frowned upon if done)	Small pebbles, rocks, soil and flowers (which in many places cannot be taken or picked) and naturally discarded items such as sheep's wool
Toys	Small-scale reproductions and models of features and objects that are about or relevant to places and activities in the environment, reproduced to be used in play activities by children	Vehicles, planes, trains, urban buildings, street furniture, farm buildings, walls and fences, animals, people and large-scale model houses and their fixtures and furniture
Replicas	Faithful and not-so-faithful copies, usually at a much-reduced scale, or stylistic representations of features and objects in the place and its local environment, usually sold as mementoes	Reproductions of well-known area-associated features, such as the Eiffel Tower or the Taj Mahal, or porcelain copies of particular local buildings

Source: Catling, 2012: 2: Varieties of Geographical Artefacts. Supplementary material

98 *Playful approaches to powerful geography*

The opportunity to handle geographical artefacts is a unique experience. Bringing mystery and drama into a geography lesson, artefacts inspire children to wonder and to ask geographical questions. Holding an artefact allows children to use their senses, and develops questioning and problem-solving skills.

Catling (2012) highlights the value of ephemeral artefacts (everyday pieces of paper intended for short-term use) including bus tickets, local maps, sweet wrappers, brochures, receipts, newspapers, all providing valuable clues about a place which can be used to support geographical learning. All children use objects on a daily basis so artefact-based enquiry (Hauf, 2010) may be less threatening to some than written sources. They create opportunities for multi-sensory exploration, facilitating entry points for learning based on children's own cultures, ethnic backgrounds, personalities and learning styles. Artefacts help make the abstract elements of culture more accessible. Particularly good for kinesthetic learners, unfamiliar objects generate curiosity and engagement. Figure 4.10 illustrates an example of clothing worn during a wedding in Malaysia.

Teachers can share their personal geographies with children through discussing stories from their own personal travels. Labbo and Field provide a structured way to do this through the compilation of a journey box. A journey box (Labbo and Field, 1999) is broadly speaking a container (e.g. box, suitcase, chest) which contains a set of selected artefacts, photographs, informational texts, maps and entries from travel journals. When combined these items tell a first-hand story of a journey in the context of place, space, time and culture. A journey box allows children to experience vicariously a journey taken by their teacher.

Figure 4.10 A Hindu wedding in Malaysia

Culture kits and artefacts from different cultures

A culture kit is a collection of artefacts or everyday objects from one specific country or region. It can bring geography lessons to life. Objects can represent both traditional and modern aspects of a culture along with secondary material such as maps and images. Authentic everyday objects (used in the household, school or workplace) demonstrate features of contemporary life in other places, including cooking utensils, tools and maps. Traditional objects may include musical instruments, clothing and images. Culture kits stimulate children's curiosity and observation. Responses to culture kits promote discussion, creativity and higher-order thinking. While the objects make the learning memorable, teachers need to design activities which provide an accurate context for the learning experience. The danger of perpetuating stereotypes and guidelines for avoiding bias are discussed in greater detail in Chapter 8.

The purpose of a culture kit is to make an unfamiliar culture more accessible to children. Focusing on similarities and differences, culture kits help children establish links between themselves and a new culture through exploration of common traits.

When using culture kits, children should have opportunities to hold, feel and describe the objects. Culture kits can enhance the study of place; however, it is important not to focus on exotic or stereotypical portrayal. Although not essential, a guest speaker or visitor from the same culture will greatly enhance exploration of the culture kit.

Sourcing culture kits

Artefacts can be acquired from a number of sources including markets, online, fair trade shops, or by collecting them or asking others to do so when visiting different places. Table 4.5 provides examples of items which can be sourced from different cultures. Items should be robust as they will be handled by several children. An amazing collection of artefacts can be assembled at low cost from teachers visiting different countries. This simply needs to be coordinated.

Becoming a geography detective: ideas for working with artefacts

A display of artefacts generates excitement in all geography classrooms. Children can work as detectives trying to gain evidence as they establish the role and function of each artefact. Working in pairs they should describe each artefact, give it a name and work out its possible use. As children engage with different objects they learn to look for meaning and to ask questions, and they begin to understand that an object can communicate many different messages. Responding to objects is a skill which needs to be developed through carefully developed activities that support questions and discussion.

What is it? (Describe the object using the information found on its label.)
What do you see? (Note the materials, decoration, size, appearance, etc.)
What can you infer? (Based on what you see, consider how this object might have been used, who might have owned it, etc.)
What more would you like to know?

The questions in Table 4.6 should help teachers maximise the educational potential from cultural artefacts.

Table 4.5 Suggestion of the kinds of artefacts that could be collected

Artefact	West and East Africa	South-East Asia	Central and South America
Drinks	Kenyan coffee	Indonesian coffee	Columbian coffee
Dolls	The Queens of Africa dolls are one of the most popular range of African dolls made to promote African Heritage. The original dolls, Nneka, Azeezah and Wuraola, represent Nigeria's three largest ethnic groups – Hausa, Igbo and Yoruba. The mission statement of the brand is to empower young Nigerian girls and other children of African descent. The dolls also come with a booklet that explains the cultures and traditions of the ethnic group that each doll represents	Pyit Taing Htaung, roughly translated to 'up-whenever-thrown', are small toys that when tossed right themselves to their original position. These dolls are meant to inspire people to continue trying when troubles come their way. Whenever people see these toys, they try to bear their suffering with a smile	Worry dolls are small (6cm high), mostly hand-made dolls that originate from Guatemala and Mexico. Created as a remedy for worrying, the tradition is to tell the doll a worry, put the doll under your pillow, and a good night's sleep will ensue. Made from pieces of wood or wire and wrapped in wool. Scraps of traditional woven fabric are used to make the doll costumes
Toys	Toy made from recycled wire. Football (made from plastic bags wrapped in string)	Five stones: This game involves throwing, picking up and catching the little pyramid-shaped bags on the back of the hand. These little cloth bags are typically filled with saga seeds, beans, rice or sand. The game progresses by increasing the number of bags you throw and catch. Trying to catch all five 'stones' is great for improving hand–eye coordination!	La piñata: This is perhaps the most internationally recognised traditional Mexican toy. It is a ceramic or cardboard container filled with sweets or treats. Blindfolded players must try to hit and break the piñata
Fabric	African cotton: The kanga is a piece of printed cotton fabric about 1.5m by 1m with a border along the four sides and a large design in the middle. Culturally significant, it is often given as a gift. The Kitenge is a larger piece of fabric used for making clothes. The Kikoy is a piece of cotton worn by men	The krama is used by people in Cambodia to do many things. It is a scarf which rural Cambodians often wear while working in the fields for sun protection, when travelling to keep dust from their eyes, when transporting heavy objects to stabilise them on their heads or when bathing in the river to wrap around their waist. Batik is a cloth that traditionally uses a wax-resist dyeing technique. Batik is one of the principal means of expression of the spiritual and cultural values of Indonesia. Common batik patterns include flowers, birds and clouds	Any item made from Alpaca wool. Alpacas are reared in Peru, Bolivia, Ecuador and Chile and are bred specially for their fibre

Musical instruments	Musical instrument such as finger piano	The Saung Gauk is an arched harp used in traditional Burmese music and is regarded as a national musical instrument of Burma
		Panpipes
Masks	Hand-carved Kenyan mask	Monkey mask: This is a miniature version of the mask which Cambodian dancers wear when they dance as the magical monkey Hanuman. Such a mask is made of paper mâché and takes about a week to fashion
		Inca, Aztec and Maya masks
Replicas/ images of buildings/ monuments	Pictures of high-rise buildings in Nairobi	Petronas Twin Towers replica: The Petronas Twin Towers are the tallest twin buildings in the world, standing at 451.9m or 1482.6ft and consisting of 88 floors. They are a defining feature of the skyline of Kuala Lumpur, the capital city of Malaysia
		Replica of Christ the Redeemer statue

Additional resources

CD of local, regional and national music
Map of specific area
Sample school book
Series of postcards
Samples of newspapers
Coins and notes (currency)
Games

Table 4.6 Sample questions for exploring artefacts

Questions about the physical characteristics of an object	Questions about the design and construction of an object	Questions about the importance and value of an object	Questions about the function of an object
What does it look, feel, smell and sound like?	What materials is it made of?	What difference did the object make to people's lives?	What is it?
How big is it?	Why were these materials chosen?	How important was the object to: the people who made it; the people who used it or owned it; people today?	Why was it made?
What shape is it?	Could different materials have been used?		How might it have been used?
What colour is it?	Is it attractive to look at?		Who might have used it?
How heavy is it?	When and where might it have been made?		What skills were needed to use it?
Does it have any marks that show us how it was made, used and cared for?	Are there any labels?	What does the object tell us about the culture of the people who owned it?	What would it have been like to use it?
What is it made of?	Does it have manufacturer's marks or decorations?		Where might it have been used?
Is it mass-produced or unique?	Was it made by hand or machine?	Is it mass-produced, rare or unique?	Might it have been used with other objects?
Is it complete or part of an object?	Is it mass-produced or unique?	Is the object financially/ sentimentally/ culturally/ historically valuable?	Has its use changed?
Is it in good condition or worn/used?	Who might have made it?		
Is it complete or are there parts missing?	Is it made in one piece or made up of different parts?		
Is it new or second-hand?	Can it be taken apart?	Is the object traditional or contemporary?	
Has it been altered, adapted, repaired or changed?	How is it put together?	In what way is the object important today?	
	How might the object work?		
	Is it decorated or plain?		
	Are there any marks/ images on the object?		
	What do these tell us about the people who made the object or owned the object, and about the culture we are studying?		

Other geographical activities using objects include:

Case study – children compile a locational case study/project based on the artefact.
Curator – children can group objects from a particular culture, organise a display and invite other children into their classroom to view.
Caption or label writing – children can write their own captions or exhibition labels, providing information about the geographical setting of the artefact.
Creative responses – children can respond to an object through creative writing, drama or art.

Benefits of using artefacts

1 Artefacts can facilitate the development of cultural understanding that may not be possible through written sources. First-hand experience of working with artefacts is less abstract for children.
2 Artefacts have the power to motivate and challenge children's thinking, conceptions and preconceptions.
3 Handling, describing and discussing artefacts allow children to take responsibility for their own learning.
4 Artefact exploration can foster children's critical and creative engagement through sustained enquiry.
5 Children can become physically involved through hands-on interaction with artefacts, which enhances their hand–eye coordination and dexterity skills. It is a multi-sensory learning experience which is enjoyable and interesting.
6 Artefacts provide opportunities for personal engagement, questioning and reflection.
7 As children engage with artefacts they develop the ability to construct ideas and thoughts.

Limitations of artefacts

1 They provide a partial view of a culture.
2 Some artefacts may be difficult to store.

Creating a class culture kit representing their place is beneficial for understanding the advantages and limitations of these resources. As children select objects which represent activities from their everyday lives, they begin to realise that artefacts tell only part of their story. Culture kits provide useful resources for teaching geography, but they are limited. It is through recognising these limits that good geographical learning can take place. Artefacts provide only a partial picture. It is important for children to realise that a single tradition does not define an entire culture.

Playing with rocks

Children have always been fascinated with rocks. They love to scramble over large boulders and throw pebbles into a river. From sand play to pebble collections, rocks contain a certain mystique for children. Learning about rocks helps children to become excited about the Earth and its composition. The outer crust of the Earth is made of rock some of which is covered by soil and water. Rock creates and shapes the Earth's landscape. It forms our magnificent mountains, shapes the deepest oceans, and provides magnificent vistas for us to enjoy. The surface of the Earth is constantly changing due to processes such as those triggered by plate tectonics and volcanic activity. Landforms are then further shaped through the processes of weathering and erosion. Weathering is the process by which rocks are chemically altered or physically broken into fragments. Erosion is the process that loosens sediments and moves them from one place to another. Learning about erosion and weathering helps children see their landscape through the eyes of a geologist. Geology is the study of the Earth's origin, structure, composition and history and the nature of the processes which have given rise to the Earth as we know it today.

104 *Playful approaches to powerful geography*

Rock is continuously being changed, rebuilt, or recycled by the forces of the Earth. Rocks come in all shapes, sizes and colours. Geologists classify rocks based on their texture and composition. Rocks are classified according to the manner of formation. One rock group is made when hot lava cools down and hardens. Another is made when sand or other small pieces get stuck together. The third rock type is made when a rock gets buried deep and heated up at an extreme temperatures. While not required for small children to know the different kinds of rocks – as igneous, sedimentary or metamorphic – 6-year-olds from Coolarne NS demonstrated high levels of rock expertise (Case study 4.1).

Primary teachers can build their own rock collection including both everyday rocks and more unique samples including those purchased or obtained from special places, such as pieces of crystal and brightly coloured minerals. Figure 4.11 illustrates student teachers working with rocks during their visit to the Burren. Each year children add their own samples (labelled carefully with their names) to enhance the class collection. Children love to touch, feel and describe their rocks. Mostly they simply love to play with rocks. Using handheld lenses, children can learn to notice minute details. Different criteria can be used for sorting, such as size, colour and type. Children can write and illustrate rock reports and a rock museum is a favourite part of the classroom. Scoil Íde, Corbally, Limerick has a permanent rockery as part of its Outdoor Classroom (Figure 4.12). By studying the world beneath our feet children learn to appreciate its impact on the world above the ground. If you are nervous about your rock knowledge begin with some of the rocks common in your locality and build your knowledge from there.

Rocks and minerals feature in our everyday lives. Children are fascinated to learn about their dependence on rocks and minerals. From large buildings and roads to cooking utensils and make-up, rocks and minerals feature prominently in our lives. Famous rocks such as Mt Rushmore, Uluru and the Rock of Gibraltar are tourist attractions. Throughout history humans have used rocks and their components to express themselves through rock paintings, chalk drawings and the creation of a pencil. Children from Coolarne NS explored the connections between rocks and a pencil. They discovered there is graphite in the lead, minerals in the paint and in some cases metal is used to connect an eraser to the pencil itself.

Figure 4.11 Student teachers working with rocks

Figure 4.12 Rockery in Scoil Íde's outdoor classroom

Rocks are used to remember our ancestors. Megalithic structures made of large stones without the use of mortar or concrete can be found in Newgrange, Ireland and Stonehenge in Wiltshire, England. Today the names of our dearly departed are proudly displayed on headstones in graveyards. Rocks have been used to create the most spectacular landmarks on the planet including the Sydney Opera House, the Washington Monument and the Taj Mahal. Rocks dominate the nature and appearance of landscape as is evident in the Burren discussed in Chapter 3. With its moon-like landscape, the Burren is one of the largest limestone areas in Europe and spreads over 250 square kilometres. The Burren, a spectacular landscape carved out of limestone slabs, facilitates the growth of the most unusual plants in Ireland. Some images from the Burren are included in the central photocollage of this book.

Making volcanoes

The best way to introduce volcanoes is with a rock. Igneous rocks are produced as a result of volcanoes. Recent eruptions around the world have highlighted the fascinating and terrifying power of volcanoes. There are some excellent YouTube clips about volcanoes including those made by *National Geographic*. Children from Gael Scoil de hÍde researched some of the world's most famous volcanoes including Mount Fuji and Vesuvius. Using latitude and longitude coordinates and working in groups, children provided a case study of their volcano.

Part of this study included making a volcano. The best type of model volcano is one that 'erupts'. To do this, the children built play dough and paper mâché volcanoes around a plastic drinks bottle. They placed each volcano on a waterproof tray. They added a tablespoon of baking soda, a tablespoon of soap powder, a few drops of red or orange food colouring and 180ml of water to the bottle, and mixed it. After adding 45ml of vinegar, the children stood back and waited for the eruption (Figure 4.13). The children compiled a glossary to describe what happens when a volcano erupts. Key terms were identified as follows: *volcano, vent, core, dormant, extinct, crust, magma, shield, lava, active, cinder, cone, crater, erupt, Vesuvius* and *Pompeii*. A card-sorting activity for teaching about volcanoes to consolidate learning and to raise further enquiry questions is available in Appendix 1.

Figure 4.13 Working with volcanoes

Case study 4.1 Rock 'n geography

As the motorway near Coolarne NS, Galway was built, layers of rock were exposed. Most of these rocks are sedimentary – formed from particles of older rocks that have been broken apart by water and wind. These particles bury living and dead plants and animals, which eventually become fossils. The class teacher collected rocks from the roadway and expanded her collection with rocks from local beaches and the Burren. Parents contributed to the rock collection, including some lovely Zebra rock from Australia. The teacher's class of 5- and 6-year-olds conducted an amazing project on rocks. Some images are included in the central photocollage of this book (see Section 6 of the colour plates).

The children held the rocks and described them. They learnt rock words. Each child painted their own rock. They also made art with chalk, which the children discovered is a type of rock. The children learnt about the three types of rocks: igneous, sedimentary and metaphoric. Their learning was reinforced through hand signs as illustrated in Figure 4.14.

Rocks formed by volcanoes are called igneous rocks (igneous is derived from the Latin for fire – *ignis*). Igneous rocks are produced as the result of volcanoes when magma or lava cools and becomes solid rock again, e.g. pumice, basalt and granite. To illustrate this rock group, the children make a sign for igniting a match. Basalt, an example of igneous rock, features in the Giant's Causeway, Co. Antrim, and is illustrated later in this book (Figure 6.5).

Sedimentary rocks are formed from particles of sand, shells, pebbles and other fragments of material. Together, all these particles are called sediment. Gradually, the sediment accumulates in layers and over a long period of time hardens into rock. Generally, sedimentary rock is fairly soft and may break apart or crumble easily. Examples of this rock type include sandstone and limestone. To illustrate this rock group, the children make a sign for crumbling material into small pieces.

Metamorphic rocks such as marble are formed under the surface of the Earth from the metamorphosis (change) that occurs due to intense heat and pressure (squeezing). To illustrate this rock group, the children make a sign for change like a butterfly.

Figure 4.14 Children from Coolarne NS demonstrating their understanding of the three main rock groups with hand signs

Using their five fingers they named different rock words: *sand, pebbles, stones, rocks* and *boulders*. A broad range of words were used by the children for describing rocks. They investigated many enquiry questions such as: How did the stone become so smooth? They examined rocks swirling around in fast-flowing water and learnt that rocks rub off each other until they are smooth.

An art teacher painted a stone illustrating the school situated on the layers of grass, soil and rock (see Section 6 of the colour plates). This is a wonderful artefact for the school which will call on viewers to consider our position on the surface of the Earth supported by layers of rocks and soil which sustain the children, their food chain and their general wellbeing.

With their own exhibition stand, the children presented their work at the Science and Technology Fair in Galway. Confidently guided by the children, members of the public were invited to identify different rocks. Children who attended had an opportunity to feel soil and sieve the sand to see shells and rocks remaining. Teachers commented on the therapeutic nature of this sensory activity.

The most spectacular part of this project was the advanced concepts addressed by the children and their sense of ease with complex terminology. This project was remarkable in terms of the level of conceptual knowledge achieved by the children. The learning also spread throughout the school community. Parents became involved and the children made presentations to other classes. The following list of true or false statements was devised by the class teacher. This list was contemplated by all children in the school.

True or false statements

Geologists investigate the Earth. They study rocks, soils and minerals.
A quarry is the place rocks are dug out of.
Granite is a hard rock that has many different colours.
Basalt is the rock found in the Giant's Causeway.
Limestone is used to make cement.
Sandstone is a soft rock and can be used to make glass.

The Burren in County Clare is made of limestone.
The Taj Mahal in India is made entirely of marble.
In the Stone Age rocks were used as tools.
Slate is the rock often used for roofing.
Chalk is a soft white rock used to write on blackboards.
Pebbles are small rocks.
Boulder is the word for a large rock.
Marble is the rock often used for statues.
Fossils are the remains or prints of ancient animals or plants.
Soil is a thin layer on the Earth's surface where plants can grow.
Soil is made of air, water, sand, clay and silt. Soil contains living and non-living things.
Common types of soil in Ireland are clay, loam, sandy soil and silt.
Earthworms tunnel through the ground bringing nutrients to the soil.
Footpaths are made of kerbstone and flagstone.
The most common creature in the soil is the earthworm.
Loam is a mixture of sandy soil and clay.
Loam is the best soil for growing most plants.
Humus is decaying plant and animal remains broken into tiny pieces.
Asteroids, meteorites and comets are space rocks.
Pottery is made from clay.
Clay feels sticky when it is wet.
Sandy soil crumbles and falls apart.
Silt is a very fine, dusty soil that is found near rivers.
The average garden has 250 earthworms per square metre of soil.
Wind, rain and living things cause rocks to break down into much smaller pieces.
Wind and water carry away pieces, which together with rotting material from dead plants and animals form soil.

Playing with rocks can be extended through other curricular areas including science and literacy. A detailed list of picture books about rocks is included at the end of this chapter.

Conclusion

Play is critically important in all of our lives. It allows us to deal with constantly changing circumstances. The literature cited in this chapter suggests that play is critical for our sense of wellbeing, successful social relationships, creativity and innovation. Through play children are more likely to take risks, and role play scenarios which ultimately help them develop lifelong skills. Play can provide a motivational strategy to engage children in geographical learning. Through play children can develop, enhance or practise specific geographical skills or engage meaningfully with geographical concepts. However, the learning needs to be supported, nurtured and reinforced by the teacher through questioning, discussion and modelling. Nevertheless, the conceptual understanding displayed by children illustrates that geographical teaching is sometimes insufficiently challenging. Children are capable of engaging with geographical concepts in a sophisticated manner in a learning context which is well supported and which has been created by a teacher with competent subject knowledge.

Playful approaches in this chapter includes children creating their own landscapes, sorting rocks and making volcanoes. Building 3-D models and playing with materials helps the process of visualisation by making abstract ideas tangible. Playing geographical games enables children to generate new and varied ideas spontaneously. The transformative potential of play has been noted by many commentators with particular reference to its ability to help children and indeed adults to be 'otherwise'. Playful approaches provide children with a chance to: explore ideas; develop language skills; try out different possibilities; predict what will happen next and cultivate their imagination. Pound (2018: 19) highlights the benefits of geographical play succinctly as follows:

> As all geographers know, people, things and ideas move, thus transforming our world. We can recognise the way in which, through their play, young children move towards becoming the creative geographers of tomorrow.

Exercise 4.1 Personal reflection for teachers

Consider the last time you had the opportunity to play. What were the circumstances and how did you feel?

How do you use play as an educational approach in the classroom?

What supports do you need to increase children's access to playful approaches in your classroom practice?

Have you ever used artefacts in your teaching? What items do you have in your possession which could potentially be used as a classroom resource for teaching geography?

On your next day trip to a nearby location collect three items which could be brought back to the classroom as geographical resources.

Further resources

Mission Explore: www.missionexplore.net. Mission Explore is an award-winning series of children's books designed by Helen Steer and illustrated by Tom Morgan-Jones. Children who complete missions from the book can visit the Mission Explore website (www.missionexplore.net) to be rewarded with points and badges for their efforts. Spearheaded by Daniel Raven-Ellison and Alan Parkinson, a team of geography teachers, academics, artists, therapists and creative individuals came together to develop the Mission Explore concept.

Askins, K. (2011) *Mission: Explore camping*. Can of Worms Kids Press.
Geography Collective (2010) *Mission: Explore*. Can of Worms Kids Press.
Geography Collective (2016) *Mission: National parks*. Can of Worms Kids Press.
Geography Collective (2010) *Mission: Explore on the road*. Can of Worms Kids Press.
Geography Collective (2012) *Mission: Food*. Can of Worms Kids Press, available to download www.johnmuirtrust.org/initiatives/mission-explore-food
Geography Collective (2013) *Mission: John Muir*. Can of Worms Kids Press, available to download www.johnmuirtrust.org/assets/000/000/849/Mission_Explore_John_Muir_original.pdf?1438098041
Geography Collective (2013) *Mission: Water*. Can of Worms Kids Press, available to download www.johnmuirtrust.org/assets/000/000/525/Mission_Explore_Water_original.pdf?1435771316

Simulation games

Banana Split game https://cafod.org.uk/content/download/733/6290/version/3/file/Primary_Fairtrade_banana-split-game.pdf
Chocolate Trading game www.christianaid.org.uk/schools/chocolate-trade-game
I Should Be So Lucky game https://cafod.org.uk/content/download/4435/36667/version/4/file/Youth_Prayer_Confirmation_I-should-be-so-lucky.pdf
Paper Bag game www.christianaid.org.uk/schools/paper-bag-game
Wake Up and Smell the Coffee game https://cafod.org.uk/content/download/855/6778/version/13/file/Wake%20up%20and%20smell%20the%20coffee.pdf
World Trading game www.christianaid.org.uk/schools/trading-game

Books about playful learning

Brown, S. (2010) *Play: How it shapes the brain, opens the imagination and invigorates the soul*. London: Penguin.
Moyles, J. (2010) *Thinking about play: Developing a reflective approach*. Milton Keynes: Open University Press.
Moyles, J. ed., (2015) *The excellence of play* (4th ed). Milton Keynes: Open University Press.

Books about artefacts

Howard, C. (2009) *Investigating artefacts in religious education: A guide for primary teachers*. UK: Religious and Moral Education Press (RMEP).
Petroski, H. (1992) *The evolution of useful things: How everyday artifacts – from forks and pins to paperclips and zippers – came to be as they are*. New York: Vintage Books.
Petroski, H. (2002) *The pencil: A history of design and circumstance*. New York: Alfred A. Knopf Incorporated.
Zuccotti, P. (2015) *Everything we touch: A 24 hour inventory of our lives*. New York: Viking.

Articles published in the Geographical Association's journal *Primary Geography*

Amswych, F. (2018) Learning through play. *Primary Geography*, *95*, 8–9.
Catling, S. (2012) The place of artefacts in geography. *Primary Geography*, *78*, 30.
Keogh, L. (2018) Clicking with role play. *Primary Geography*, *95*, 20–21.
Morse, C. and Witt, S. (2014) Go, be playful. *Primary Geography*, *84*, 16–17.
Murray, J. (2018) Discovering play. *Primary Geography*, *95*, 6–7.
Nichols, S.A. (2018) Playful transitioning. *Primary Geography*, *95*, 10–14.
Parkinson, A. (2018) Pokémon gone. *Primary Geography*, *95*, 15.
Passananti, G. (2018) Bringing environments to life though play. *Primary Geography*, *95*, 24.
Pike, S. (2018) Geographies of play, hazards and risk. *Primary Geography*, *95*, 32.
Pound, L. (2018) Playing to understand the world. *Primary Geography*, *95*, 18–19.
Robertson, J. (2018) Making playful connections. *Primary Geography*, *95*, 28–29.
Whittle, J. (2018) The potential for play on fieldtrips. *Primary Geography*, *95*, 26–27.
Witt, S. and Clarke, H. (2016) Making connections in the company of pigeons. *Primary Geography*, *91*, 12–14.
Witt, S. and Clarke, H. (2018) There's no play like gnome. *Primary Geography*, *95*, 10–14.
Young, J. (2018) Playing with place. *Primary Geography*, *95*, 22–23.

Picture books about rocks, soil and the world beneath our feet

Christain, P. and Hirsch Lember, B. *(Photographer)* (2008) *If you find a rock*. Boston, MA: Houghton Mifflin.
Claybourne, A. and Garland, S. (illus) (2017) *This little pebble*. London: Franklin Watts.
Guilliane, C. and Zommer, Y. (illus) (2017) *The street beneath my feet*. London: Words and Pictures.
Hegarty, P. and Clulow, H. (2017) *Above and below*. London: Little Tiger Press.
Hooper, M. and Coady, C. (illus) (2015) *The pebble in my pocket: A history of our earth*. London: Frances Lincoln Children's Books.
Hutts Aston, D and Long, S. (illus) (2015) *A rock is lively*. San Francisco, CA: Chronicle Books.
Hyde, N. (2011) *How to be a rock collector*. St. Catharines, ON: Crabtree Publishing.
Manning, M. and Granström (2014) *What's under the bed? A book about the Earth beneath us*. London: Franklin Watts.
Mc Guirk, L. (2011) *If rocks could sing: A discovered alphabet*. Toronto, ON: Tricycle Press.
Mizielinski, A. and Mizielinski, D. (illus) (2016) *Under earth under water*. London: Templar Publishing.
Purdie Salas, L. and Dabija, V. (2015) *A rock can be…* Minneapolis, MN: Millbrook Press.

References

Ackermann, E. (2014) *Amusement, delight, whimsy, and wit, the place of humor in human creativity*. Paper read at Constructionism 2014 International Conference, Vienna, Austria, August.
Askins, K. and Raven-Ellison, D. (2012) Spotlight on … mission: Explore food. *Geography*, 97, 163.
Barnes, J. (2003) Creating the world in the mind. *Primary Geographer*, 50, 16–18.
Bruce, T. ed., (2012) *Early childhood practice: Froebel today*. London: Sage Publications.
Catling, S. (2012) The place of artefacts in geography. *Primary Geography*, 78, 30.
Catron, C.E. and Allen, J. (2007) *Early childhood curriculum: A creative play model*. New Jersey: Pearson-Merrill Prentice Hall.
Chawla, L. ed., (2016) *Growing up in an urbanizing world*. London: Routledge.
Clemens, R., Parr, K. and Wilkinson, M. (2013) Using geographical games to investigate 'our place'. *Teaching Geography*, 38(2), 63–65.
Csikszentmihalyi, M. (1997) *Finding flow the psychology of engagement with everyday life*. New York: Basic Books.
Early Childhood Ireland (2015) Small World Play available on www.earlychildhoodireland.ie/small-world-play/
Greene, M. (1995) *Releasing the imagination: Essays on education, the arts, and social change*. San Francisco: Jossey-Bass.
Hauf, J.E. (2010) Teaching world cultures through artifacts. *Journal of Geography*, 109(3), 113–123.
Hicks, D. (2014) *Educating for hope in troubled times: Climate change and the transition to a post-carbon future*. London: Trentham Books Limited.
Hoodless, P. (2008) *Teaching history in primary schools*. Exeter: Learning Matters.
Kafai, Y.B. (2006) Playing and making games for learning instructionist and constructionist perspectives for game studies. *Games and Culture*, 1(1), 36–40.
Kalvaitis, D. and Monhardt, R.M. (2012) The architecture of children's relationships with nature: a phenomenographic investigation seen through drawings and written narratives of elementary students. *Environmental Education Research*, 18(2), 209–227.
Kress, G. (2010) *Multimodality. A social-semiotic approach to contemporary communication*. Oxon: Routledge.
Labbo, L.D. and Field, S.L. (1999) Journey boxes: Telling the story of place, time, and culture with photographs, literature, and artifacts. *The Social Studies*, 90(4), 177–182.

Louv, R. (2010) *Last child in the woods: Saving our children from nature-deficit disorder.* London: Atlantic Books, Limited.

Louv, R. (2013) *The nature principle: Human restoration and the end of nature-deficit disorder.* Chapel Hill, NC: Algonquin Books.

Morse, C. and Witt, S. (2014) Go, be playful. *Primary Geography, 84,* 16–17.

Moyles, J. (2010) *Thinking about play: Developing a reflective approach.* Milton Keynes: Open University Press.

NCCA (2013) Learning and developing through play. www.ncca.biz/Aistear/pdfs/Guidelines_ENG/Play_ENG.pdf

Pacini-Ketchabaw, V., Kind, S. and Kocher, L.L.M. (2017) *Encounters with materials in early childhood education.* Abington: Routledge.

Parkinson, A. (2009) Think inside the box: Miniature landscapes. *Teaching Geography, 34*(3), 120–121.

Skovbjerg, H.M. (2018) Theories about trampoline jumping and playing catch are going to create good design. Available on www.designskolenkolding.dk/en/news/theories-about-trampoline-jumping-and-playing-catch-are-going-create-good-design

Slinkachu (2008) *Little people in the city: The street art of Slinkachu.* London: Boxtree Publishers.

Tidmarsh, C. (2009) Using games in geography think piece. www.geography.org.uk/download/ga_prgtiptidmarshactivities.pdf (accessed 2 February 2018).

Walford, R. (2007) *Using games in school geography.* Cambridge: Chris Kington Publishing.

Whyte, T. (2017) Fun and games in geography. In Scoffham, S. ed., *Teaching geography creatively,* London: Routledge, 12–29.

Witt, S. (2017) Playful approaches to learning out of doors. In Scoffham, S. ed., *Teaching geography creatively,* Abingdon: Routledge, 44–57.

Woodyer, T. (2012) Ludic geographies: Not merely child's play. *Geography Compass, 6*(6), 313–326.

Woolliscroft, J. and Widdowson, J. eds., (2010) *GCSE geography OCR A.* Oxford: Oxford University Press.

Section 1 Exploring place through field trips

Section 2 Shaping our place

Section 3 Big issues in geography

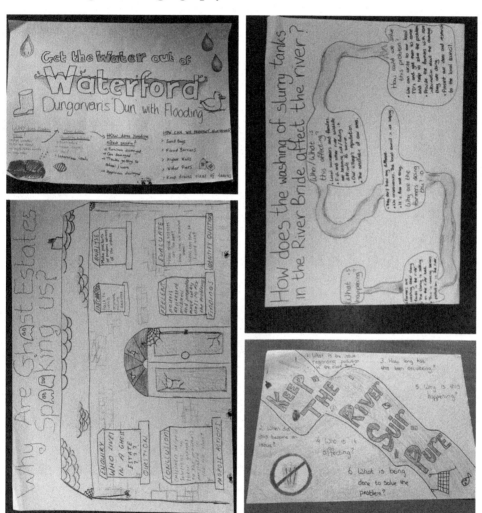

Section 4 Big day out for the Bees part 1

Section 5 Big day out for the Bees part 2

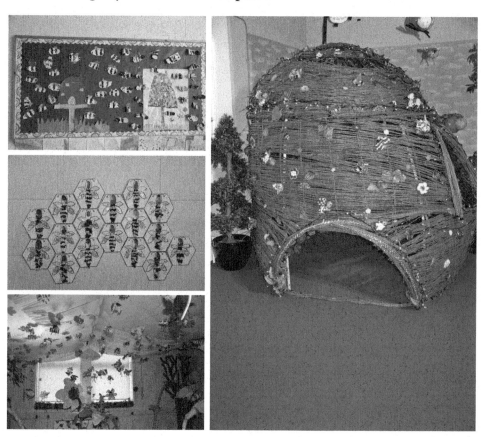

Section 6 Rock 'n' geography

Section 7 The Keep on Track Project

Section 8 Geography on display

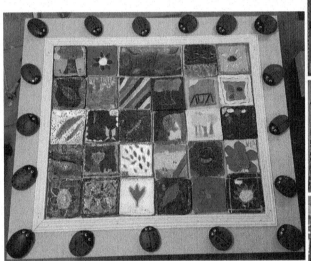

5 Teaching powerful geography through topics

Weather and climate change

Introduction

Weather and climate have a direct impact on our daily lives. The kinds of houses we live in, the kind of food we eat, the clothes we wear, our leisure activities, and even the nature of our moods, all depend on weather. A daily topic of conversation in many homes, weather is all pervasive. From a geographical perspective, understanding weather and climate processes has been described as 'an essential life skill' (Lane and Catling, 2016: 115). Climate change has been defined as the greatest global crisis of our times. Yet, the education sector remains underutilised as a strategic resource to mitigate and adapt to climate change (Mochizuki and Bryan, 2015).

This chapter sets out to:

- outline approaches to teaching weather and seasonal changes;
- highlight the importance of climate change and climate justice education;
- illustrate some of the innovative climate change education currently taking place in primary schools.

Teaching weather and climate change

Weather refers to changes in the atmosphere over a short-term period (minutes to months) whereas climate refers to changes in long-term averages of weather patterns. The word 'climate' comes from the Greek *klima* meaning 'region' or 'zone', from the old French term *climat* denoting 'region' or 'part of the earth', and from Latin *clima* indicating a 'region' or 'slope of the earth from equator to pole'. Ancient geographers divided the Earth into zones based on the angle of the sun on the slope of the Earth's surface and the length of daylight. As the change of temperature was later considered more important, the word 'climate' was associated with long-term averages of weather patterns and distinct geographical regions.

Weather is something children can easily relate to and understand. Making observations about the weather and discussing climate can facilitate children's natural curiosity. Whether it is raining or sunny, weather has an impact on children's lives every day. Rooney (2016) highlights the significance of children's relationship with weather and argues that educators need to engage with children's weather relations. Instead of learning about the weather children need to learn *in* and *with* the weather.

Children are enchanted by the weather. Being scared of thunderstorms and fascinated by rainbows are common childhood experiences. Children love to experiment with weather in their play by jumping in and out of puddles, cracking ice sheets and playing with snow. Whether it's foggy or clear, sunny or rainy, windy or calm outside, weather and climate affect children's lives every day. Children relate naturally to weather. Therefore, the potential to make observations and investigate the weather are immense. However, in some cases forecasting the weather has become synonymous with looking at a TV screen or weather apps on smartphones rather than going outside to look for signs in the sky (Gough and Gough, 2016).

Lane and Catling (2016) cite three important reasons why pre-service primary teachers should understand weather processes and climate conditions. First, it is considered part of being a well-informed global citizen. Second, it is a core part of the primary curriculum in the UK, Ireland and elsewhere. Teachers need to be sufficiently knowledgeable in order to be able to teach these topics competently and answer children's questions. Third, teachers need to be able to address children's alternative conceptions and misunderstandings. These misconnections serve as valuable starting points for discussion and concept exploration.

Weather fieldwork

As the weather is constantly changing, weather-related investigations are relatively simple to instigate. Given the unpredictability of weather in Ireland and the UK it is important for children to have rainproof clothing, wellies and a change of clothing in school. The Aistear programme (NCCA, 2013) recommends that teachers should have (1) rainy day; (2) windy day; and (3) sunny day boxes available in school in order to help children respond spontaneously to the weather. A windy day box might include bubbles, streamers, chimes, windmills, a kite and relevant picture books. A rainy day box may well include toy boats, toy ducks, umbrellas, a sieve, containers for measuring rainfall, funnels, containers for gathering water, charts for recording rainfall and relevant picture books. A sunny day box could possibly include a picnic basket, rug, teddy bears, sunglasses, sun protection cream and relevant picture books. A list of weather-related picture books is provided at the end of this chapter.

Directly experiencing the weather and observing changes that occur throughout the seasons is a fundamental part of weather and climate education. Seasonal activities are suggested in Table 5.1. There are a number of picture books about seasons, cited at the end of this chapter, which can complement children's work in their seasonal investigations.

Seasons are tied to the climate. Children from Scoil Íde, Corbally, Limerick explore seasonal developments every term. Third class children (8–9 years) monitor the movement of swallows, to see when they arrive and when they leave. In the school children are involved in all-year seasonal investigations and observations. Each class adopts a tree for a year:

- Junior infants (4–5 years): the oak tree;
- Senior infants (5–6 years): the maple tree;
- First class (6–7 years): the beech tree;
- Second class (7–8 years): the hawthorn tree;
- Third class (8–9 years): the birch tree;
- Fourth class (9–10 years): the cherry tree;
- Fifth class (10–11 years): the oak tree;
- Sixth class (11–13 years): the maple tree.

Table 5.1 Activities for exploring seasonal change

Other seasonal activities

Enquiry question: What colour is autumn? Take a leaf walk. During autumn bring the children outside for a leaf walk. Children can collect leaves and sort them according to colour, shape and size. Artwork can be created based on real-life observations.

Enquiry question: How does autumn feel? Walk around your local area. As you walk, stop and watch, listen, smell and touch. Ask the children to describe the feeling of their local area. Expand children's vocabulary by asking questions: How does the air feel? Do you feel rain, hail or mist? How would you describe the temperature? Gather material for a sensory re-creation in the classroom.

Enquiry question: What would autumn be like if she was a person? Create a seasonal character. Take a stick on a walk. Bring the children on a hunt for sticks. Encourage them to find sticks and describe them and create characters based on their sticks.

Enquiry question: How is the weather changing? Using a dedicated calendar in the classroom ask the children to track the weather each day. Younger children can design symbols for each type of weather. At the end of the month count the number of cloudy, sunny and windy days, determining the pattern and any unusual events during each season. Weather diaries are common features in many classrooms. However, while day-to-day observation may be popular, real learning occurs with the study of long-term trends.

Enquiry question: What is the sound of this season? Go outside and record seasonal sounds, such as whirling wind, crunching leaves and the sound of rain. Try to re-create these sounds in the classroom with various percussion instruments, some of which are made from seasonal materials.

Enquiry question: How does this season taste? Children can be invited to bring in some seasonal produce. What kind of foods can you eat during each season? Which season is the most delicious? Make a recipe book, complete with photos of the food you made, to document your tasty year.

Throughout the year they observe seasonal changes and take care of their adopted tree. To celebrate Tree Day on 9 October and Tree Week in March all the children spend some time with their adopted tree. Children make bark and leaf rubbings, paint tree pictures and go orienteering through the trees. Some classes go on a leaf hunt! By the time the children leave school they have an in-depth conceptual and experiential knowledge of trees based on their study of eight different varieties.

'Signs of autumn' Twitter project

Many classes share their work through a Twitter account. This provides the school with an opportunity to share its work and to receive feedback from other schools. One such initiative is the Signs of Autumn Twitter Project established by Seomra Ranga. Classes are invited to report any signs of autumn observed in the school area or locality through Twitter. Using the hashtag: *#anfomhar* (Irish word for 'autumn') classes tweet a photo of the sign of autumn noticed or observed. This technology allows children to connect with other classes across the country and internationally. Classes are encouraged to reply to, comment on or re-tweet to promote dialogue, connectivity and collaboration.

Signs of autumn recorded are as follows:

- Leaves falling from the trees;
- Bare autumn trees;

116 *Teaching powerful geography through topics*

- Windy autumn weather;
- Autumn colours seen in the local environment;
- Seasonal fruit/vegetables;
- Animals spotted in the locality, e.g. hedgehogs;
- Children playing conkers;
- Children's autumn artwork/display.

The Greenwave Project was an initiative for primary children to observe and track the arrival of spring as it moves across Ireland; see Figure 5.1. Every year a green wave caused by the opening of buds on trees and hedges can be seen from outer space moving across Europe in springtime; hence the project's name. This green wave begins in the south of Europe in February and moves up across the continent as temperatures rise. The phenomenon travels at about the same speed as humans walk, 6.4km per hour. Based on that speed, spring takes three weeks to move across Ireland from Mizen Head to Malin Head.

Children from schools all over Ireland were invited to log their sightings of six species: ash, horse chestnut and hawthorn trees, the primrose, the swallow and frogspawn on a project website. They were also asked to measure and record wind speed, rainfall and temperature. Children (11–13 years) from Scoil Íde in Limerick continue to monitor the signs of spring on an annual basis. As the days grow longer and temperatures rise children are assigned to record their observations every day before school begins. By completing this process over a number of years the children are able to draw comparisons and notice trends, as illustrated in Table 5.2.

Figure 5.1 Greenwave: Signs of Spring, Scoil Íde, Corbally, Limerick

Table 5.2 Children's records of the signs of spring gathered on an annual basis

	Frogspawn	Hawthorn tree	Horse Chestnut tree	Primrose	Ash tree	Swallow
2015	5 March	10 March	2 March	22 April	10 April	30 April
2016	22 February	16 March–3 April	16 March	6 April	12 April	18 April
2017	1 March	15 March	20 March	24 April	14–24 April	9 April
2018	27 February	27 March	7 April	20 March	19 April	30 April

Forecasting the weather: making instruments to measure the weather

Meteorologists study the weather using a variety of technical equipment to monitor changes in the weather over time and make predications for the future, e.g. a weather forecast. A weather forecast can include information on precipitation (rain, snow, sleet or hail), wind (direction and strength) and air pressure. Ideally, every school should have a weather station. This is a facility, with instruments and equipment for measuring temperature, atmospheric pressure, humidity, wind speed, wind direction and precipitation amounts. This equipment is set out in Table 5.3. Instructions for making your own weather equipment are widely available online and a home-made anemometer is illustrated in Figure 5.2. An important part of weather forecasting is the collection and recording of data. Figure 5.3 illustrates one school's records of daily temperatures during the month of December.

Table 5.3 Instruments for measuring the weather

Instrument for measuring weather	Description
Rain gauge	A rain gauge measures how much rain has fallen
Thermometer	A thermometer measures temperature. Track and note the temperature throughout the day
Barometer	A barometer tracks the air pressure. If the pressure drops that often means clouds and precipitation (rain or snow). If the pressure goes up that usually means clear or fair weather
Anemometer	An anemometer measures wind speed
Weather vane	A weather vane tells you which way the wind is blowing
Hygrometer	A hygrometer measures the humidity, the moisture in the air

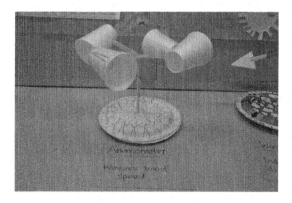

Figure 5.2 An anemometer designed by a 9-year-old child

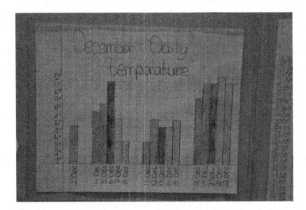

Figure 5.3 Weather data collected by the children

In line with enquiry-based learning, measuring the weather is more effective if it is conducted as part of an investigation which is of interest to the children. The enquiry question *Is playground behaviour influenced by the weather?* was investigated by children in St Anne's NS, Shankill, Dublin. Children investigated the impact of the weather on behaviour in their playground. Over the month of December they took detailed daily measurements of the weather including wind speed and temperatures. The children recorded: the number of disagreements; requests for drinks and/or coats; and a visit to the toilet. Their findings, along with details of their enquiry, were presented to their school community and to a national audience during a national primary education science festival (Figure 5.4).

Figure 5.4 Presentation of children's findings

Broadcasting the weather forecast: The vocabulary of weather

Recording and broadcasting the weather are popular activities in classrooms. In order to do this purposefully and effectively, children require a good command of weather language such as those terms set out in Table 5.4. An opportunity to use and hear these terms in classroom discussions, oral presentations, drama, debate and conversation is also required.

There are several excellent resources available for teaching weather and climate. The Royal Meteorological Society (metlink.org) and Met Éireann (met.ie) provide wonderful resources for teaching some of the complexities of weather terms and concepts. Irish meteorologist Joanna Donnelly (2018) has recently published a book for primary children *The Great Irish Weather Book* (Figure 5.5). Appendix 2, containing an invaluable weather glossary for teachers, has been compiled by Paula Owens (2018).

Teaching unusual weather conditions and natural disasters

Hurricanes and typhoons are the same weather phenomenon: tropical cyclones. According to the National Ocean Service website: 'A tropical cyclone is a generic term used by meteorologists to describe a rotating, organised system of clouds and thunderstorms that originates over tropical or subtropical waters and has closed, low-level circulation' (oceanservice.noaa.gov). In 2017 hurricanes Harvey, Irma and Maria devastated the Caribbean and caused serious disruption for Texas and Florida. In the same year Cyclone Mora killed six people in Bangladesh and displaced 500,000 more. While Atlantic hurricanes tend to get more media coverage than Pacific cyclones both are linked to the same warmer waters. The same kind of phenomenon occurring in the Northwest Pacific is called a typhoon. Hurricanes are typically associated with the Caribbean and Northern America. However, in 2017 the British Isles experienced the effects of a hurricane. Ophelia originated in unusually warm tropical Atlantic waters in an area where no hurricanes normally develop and moved towards Ireland and Great Britain, generating high waves and wind gusts of more than 150km/hour.

The winds gathered hot air from the Sahara, causing autumn temperatures in Europe to become unseasonably high. Heat, wind and dry soil from extended drought fuelled fires in northern Portugal and Spain. Smoke from the fires and hot Saharan dust then spread

Table 5.4 Sample of weather words and terms

Climate	Mist	Low- and high-pressure system
Weather	Weather station	Beaufort wind scale
Drought	Weather satellite	Scattered showers
Flooding	Forecast	Tropic of Cancer
Fog	Meteorology	Westerlies
Humidity	Monsoon	Temperature
Hurricane	Hurricane	Cloudy
Meteorology	El Niño	Cold front
North/north-east/north-west	South/south-east/south-west	East and west
Weather forecast	Isobars	

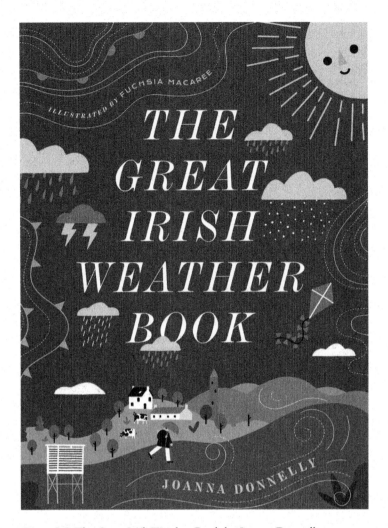

Figure 5.5 The Great Irish Weather Book by Joanna Donnelly

in the atmosphere changing the skyline in many areas, including London, where dramatic colours were reported.

The storm disrupted transport and energy systems and damaged coast and urban infrastructure. In addition to the casualties from the storm and wildfires, there were other unreported health problems such as respiratory issues. Accurate forecasts and close coordination between meteorological, marine and disaster management services limited casualties in Ireland, but several people died because of wildfires in Spain and Portugal. Other recent hurricanes show the domino effect of weather and weather-related hazards on all sectors of society.

Advances in weather forecasting mean that weather patterns can be predicted for up to six days. This means we have more advance warning of weather hazards than before, allowing people to prepare, thus reducing potential loss of life and damage to property. Providing children with weather information gives teachers the opportunity to talk with children about the importance of being on alert when there's a chance of bad weather and knowing what to do if a severe storm or natural disaster occurs. Lane and Catling (2016: 208) suggest that young adults may not have a coherent knowledge of the causes, processes and impacts of hurricanes. Their study suggests that many prospective primary teachers 'remain ill-informed about tropical cyclones as an example of weather and climate hazards'. This means they are unlikely to develop in children sound understanding of these concepts. Unusual weather events provide teachers with a perfect opportunity for exploring climate concepts and processes (Scoffham, 2018; Hatwood, 2018).

Case study 5.1 Making the link between unusual weather events and climate change

Following the arrival of Hurricane Ophelia, a number of teaching activities were devised using resources from local and national newspaper and media sources. The first activity is a mystery based on the question: *Why was Emily not invited to Laura's birthday party?* Information for the mystery cards was sourced from newspapers and news coverage of the hurricane, as shown in Table 5.5. The mystery cards can be adapted for any unusual weather event. As described in Chapter 2, the mystery cards are distributed to the children. The children read the cards, sort them into groups and solve the problem, giving reasons for their answer. This is an effective way of introducing the topic of hurricanes or revising work which has been covered to date.

Table 5.5 Mystery activity cards: Why was Emily not invited to Laura's birthday party?

Hurricanes are substantial tropical storms that produce heavy rainfall and strong winds.	Emily's mother is concerned about her daughter's overuse of her iPad.
Hurricanes form over warm ocean waters near the equator. The warm, moist air above the ocean surface rises, causing air from surrounding areas to be 'sucked' in. This 'new' air then becomes warm and moist and rises too, beginning a continuous cycle that forms clouds. The clouds then rotate with the spin of the Earth. If there is enough warm water to feed the storm, a hurricane forms!	Emily keeps in touch with her friends through Snapchat and Instagram.
Most hurricanes occur harmlessly out at sea. However, when they move towards land they can be incredibly dangerous and cause serious damage.	Emily is 10 years old and her friend Laura is planning a big birthday party. She is planning to have a huge bonfire and her parents have bought additional supplies of coal and turf for this event. She is inviting her friends through Snapchat and Instagram.
Ireland and the UK Met Offices started naming all storms in 2016. Hurricane Ophelia came from the USA; therefore it was given its name by the World Meteorological Organization (WMO). Storms are named to raise public awareness of weather events.	Researchers have found that the average 8–10-year-old child spends nearly eight hours a day with different media during holiday time.

(Continued)

Table 5.5 (Continued)

Thousands of customers in Cork, Wexford, Limerick, Tipperary and Kerry were left without broadband and mobile services after Hurricane Ophelia. Thousands were left without electricity or water.	Emily and Laura have been best friends since they met at Montessori. However, they had a big row the week before the birthday party.
Researchers believe that the amount of time children spend on smartphones and other technological devices should be lowered as it can lead to obesity, lack of sleep and school problems.	Hurricanes rotate around a circular centre called the 'eye', where it is generally calm with no clouds. Surrounding the eye is the eye wall – the most dangerous part of the hurricane with the strongest winds, thickest clouds and heaviest rain!
A warmer ocean surface means more moisture in the atmosphere.	Emily's father works as a pilot for British Airways.
Hurricanes are also called cyclones and typhoons, depending on where they occur.	Laura lives in the coastal village of Cobh, Co. Cork. Her house was flooded due to a storm surge. When a hurricane reaches land it often produces a storm surge. This is when the high winds drive the sea towards the shore, causing water levels to rise and creating large crashing waves.
Sea level rise has contributed to the coastal flooding associated with recent major hurricanes.	Scientific evidence proves that hurricanes are getting stronger.
Hurricane Ophelia knocked a number of trees outside Emily's house. As a result her house was damaged and the road to her house was blocked and cars could not travel on this road. The strong spiralling winds of a hurricane can reach speeds of up to 320km/hr – strong enough to knock over trees and destroy buildings!	Storm Ophelia rotated in an anti-clockwise direction. In the southern hemisphere, hurricanes rotate in a clockwise direction, and in the northern hemisphere they rotate in an anti-clockwise direction. This is due to what's called the Coriolis force, produced by the Earth's rotation.

After conducting this mystery children are asked to identify impacts of the hurricane. Small cards are distributed whereby they record their ideas, which are then placed under one of four headings: *social, economic, political* and *natural*. The cards from Table 5.6 are then distributed and children are asked to identify which of the headings they belong to. The final result is set out in the shape of a large compass, which facilitates further discussion as shown in Figure 5.6.

Table 5.6 Impacts of Hurricane Ophelia in Ireland

Social	Economic	Political	Natural
Three people tragically lost their lives.	Insurance companies estimate the cost of the storm was €1 billion.	Red warning alert issued by Met Éireann and supported by the Government.	400 trees were knocked over in Cork city and county.
Over 400,000 homes were left without electricity.	Several businesses were left without electricity.	The Minister for Education ordered all schools to close for two days. Schools were closed across the Republic of Ireland and Northern Ireland for two days.	Very strong wind speeds were measured off the coast of Cork.

Many homes were left without water.	The local coffee shop couldn't open because there was no electricity.	Ireland will fail to meet its climate target of an 80% emissions reduction by 2050.	Ophelia created the biggest waves ever recorded off the Irish coast.
Roof tiles were loosened by the storm.	Loss of power, telephone and internet.	Is the Irish Government committed to its promise to reduce greenhouse gas emissions by 2030?	Brian couldn't attend the gym because the road from his house was blocked with trees.
Aileen couldn't take a shower because there was no water in her house.	Aer Lingus and Ryanair cancelled flights.	A childminder was fired by her employer because she refused to take young children shoe shopping.	The roof of Douglas Community School's PE hall went flying through the air.
Amy's trampoline was blown away.	John works in a bar. He lost two days' pay because the bar had to close down because of the damage caused by the storm.	Due to disruption caused by Hurricane Ophelia the application deadline for a climate change action competition was extended.	Ophelia was a rare Category-3 hurricane.

Figure 5.6 A card-sorting activity

The children are then divided into groups to discuss the impact of Hurricane Ophelia from one of the perspectives of: insurance company, climate expert, politician and person affected by the disaster. Children are asked to review their learning so far and to make three important points about the hurricane from their given perspective.

Card 1 Insurance company

After Hurricane Ophelia you are going to face a significant increase in claims from people who have experienced damage. How are you going to deal with these claims and what priorities are you going to propose?

Card 2 Experts from Met Éireann

According to your evidence, more hurricanes will reach Ireland in future years. What are your proposals to the Irish public in terms of dealing with these storms? Please use some of your expertise on climate change in your presentation.

Card 3 Laura's family

Your home has been flooded. How are you going to plan your claim? How will you persuade the insurance company that your claim should receive maximum compensation?

Card 4 Emily's family

Your family home has been badly damaged by trees. How are you going to plan your claim? How will you persuade the insurance company that your claim should receive maximum compensation?

Card 5 Local politician

While you are extremely sympathetic you are unsure how to deal with the current situation. You will need votes from Emily's and Laura's families in the next election so you must be very careful. Please prepare a presentation outlining some actions that you will take.

Once the groups have prepared their points the teacher conducts an interview as a reporter from the National News Station. The children speak in role and interviews are recorded on video and replayed to the class. Using their school web page, children from Clontuskert NS uploaded their video as part of their reporting on climate change (https://clontuskert.scoilnet.ie/blog).

Climate change

Climate change caused by global warming is already beginning to transform life on Earth. It is the defining challenge of our time, perhaps the most significant challenge facing all citizens today. The evidence of climate change is compelling. There is widespread consensus among the international scientific community that human-induced climate change is happening. According to the United Nations (UN),

> greenhouse gas emissions from human activities are driving climate change and continue to rise. They are now at their highest levels in history. Without action, the world's average surface temperature is projected to rise over the 21st century and is likely to surpass 3 degrees Celsius this century.
>
> (UN, 2017)

It is time to accept individual and collective responsibility.

Earth-orbiting satellites and other technological advances, collecting many different types of information about our planet and its climate on a global scale, have enabled scientists to see the big picture. This body of data, collected over many years, reveals the signals of a changing climate. The planet's average surface temperature has risen about 2.0 degrees Fahrenheit (1.1 degrees Celsius) since the late 19th century, a change driven largely by the release of carbon dioxide and other human-made emissions into the atmosphere. The ten hottest years on record have occurred since 1998. Rising sea levels, declining Arctic sea ice, changes in precipitation patterns resulting in extreme flooding, droughts, and more extreme weather events such as heatwaves, cyclones and tropical storms are just some of the effects of changes to the global climate. Other impacts include increased acidification and warming of the oceans, decreased snow cover, glacial retreats and shrinking ice sheets. Each of these changes is resulting in serious knock-on effects such as increased poverty, species extinction, conflict and migration. The threats to biodiversity and species are unnerving. It is alarming to realise that climate change is no longer just about polar bears and penguins; it is not only about coral reefs and sea turtles; it is about our survival on planet Earth.

The terms 'mitigation' and 'adaptation' refer to two different avenues for dealing with the challenges raised by climate change. *Mitigation* deals with the *causes* of climate change and aims to reduce the amount of greenhouse gases released into the atmosphere. Alternatively, *adaptation* involves making changes to prepare for and negate the *effects* of climate change, thereby reducing the vulnerability of communities and ecosystems. By adapting to cope with the effects of climate change, communities, enterprises and institutions can build up their adaptive capacity and *resilience* (Krasny and DuBois, 2019).

The transition to a low-carbon, climate-resilient economy has to be part of the solution as it makes economic, social and environmental sense (Hicks, 2014). Political will and leadership are needed for this to happen. There is also a need to think intergenerationally. We need to imagine the world of 2050, 2060 and 2070, with an anticipated population of 9 billion people, and take the right decisions now to ensure that our children and grandchildren inherit a liveable world (Tutu and Robinson, 2011).

Climate change denial

There has been widespread reluctance to comprehensively address the climate change threat, to the extent that 'denial has become a ubiquitous response to climate change' (Kagawa and Selby, 2010: 42). Facing up to climate change also means confronting the uncomfortable reality that the growth-based economic and political models on which we depend are complicit in the problem. Despite the need for immediate action, an increasingly remote global elite stall negotiations and expand their plans for carbon-based resource exploitation. Climate deniers – those who wish to maintain the status quo – include those who have benefited from wealth created either directly or indirectly from greenhouse gases. The election of President Trump in the United States of America signals the success of these business interests and the indifference of voters to climate change.

In addition to denial there are also widespread misconceptions about climate change which need to be challenged especially among populations of affluent societies. Education too has been slow to take on board the challenges raised by climate change.

According to Kagawa and Selby (2010: 5), 'the academy has tended to fiddle while Rome burns'. There has been little evidence of learners engaged in openly debating and discussing the causes, personal responses and societal implications of climate change scenarios that are likely to play out during their lifetimes. In other words, climate change learning experiences have tended to remain within the 'business as usual' approach (Waldron et al., 2019: 4).

Misconceptions about climate change with secondary students include confusion about the difference between weather and climate, lack of understanding about the greenhouse effect, and a belief that climate change is caused by pollution or ozone holes

(Choi et al., 2010). In a study from Western Australia only one in three high school students from Year 10 was able to write correctly or partially correctly a definition of the greenhouse effect and climate change (Dawson, 2015). In some cases misconceptions were fuelled by incomplete information from teachers (Tsaparlis, 2003).

Research indicates that pre-service teachers also have misconceptions about the causes and consequences of climate change. These include a mistaken belief that there is a relationship between the greenhouse effect and ozone layer depletion, and misunderstandings about the difference between the terms 'climate change' and 'weather' (Papadimitriou, 2004; Lambert et al., 2012). Boon (2016) found that, even in a university with a specialist focus on sustainability education, pre-service teachers had difficulties with the scientific complexity of climate change.

A climate justice response

In 2015, at COP21 in Paris, otherwise known as the Paris Climate Conference, 188 countries signed a legally binding agreement to maintain global warming below 2 degrees Centigrade; this agreement in now in effect. This represents a coherent international response to climate change. The injustice of climate change continues to be raised by campaigners and non-governmental organisations (NGOs) including Trócaire, Christian Aid and Oxfam. Calling for a climate justice response, the NGOs recognise that people who have contributed least to the problem are most affected (Waldron et al., 2019). While everyone is vulnerable the impact is far greater on those in low-income countries. Those who have contributed least to the problem, people in the Global South, face the worst consequences of climate change, and are struggling to cope with drought, storms and floods, as indicated in Figure 5.7. This diagram illustrates the causes and effects of climate change and their disproportional impact on poorer countries.

Poverty and food security cannot be tackled without addressing the issue of climate change and helping people to adapt to its impacts. Climate justice links human rights and development to achieve a human-centred approach whereby the rights of the most vulnerable are safeguarded and the burdens and benefits of climate change are shared by all. Climate justice begins at home; it begins with each decision we make in relation to energy, transport and lifestyle.

Figure 5.8 illustrates Ireland's CO_2 emissions in comparison with some low-contributing countries.

Figure 5.7 Climate change hits the poor hardest

Source: Trócaire (2017)

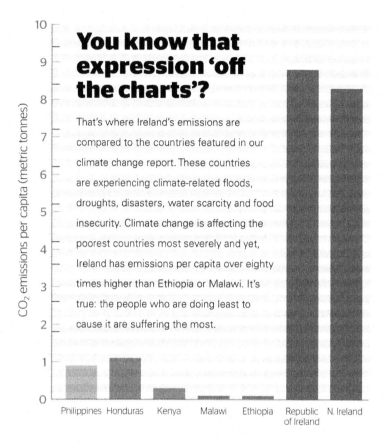

Figure 5.8 Ireland's CO_2 emissions per capita (metric tonnes)
Source: Trócaire (2017)

Climate change education

Our generation and future generations will have to learn how to live with the challenges posed by climate change. Nevertheless, the education sector 'offers a currently untapped opportunity to combat climate change' (Anderson, 2012: 193). This is due to the absence of climate change education on the curriculum and as part of the learning agenda in schools. Kagawa and Selby (2010) call for a transformative educational agenda. Climate change should be understood as a complex geographical and social as well as scientific issue, epitomised by uncertain and context-specific knowledge. This requires teachers to engage in enquiry-based learning and co-learning with children (Stevenson et al., 2017).

Mainstreaming climate change education as part of formal education systems has to be one of the most important and effective means of developing capacities for addressing climate change (Mochizuki and Bryan, 2015). Hence, climate change should feature explicitly on the curriculum. Yet, in an analysis of 78 national curricula, the curricula of half of countries worldwide do not explicitly mention climate change or environmental sustainability in their content (UNESCO, 2016). Nations that are doing well with respect to national climate change education efforts include India, Costa Rica, the Dominican

Republic, Mauritius, the Cook Islands, Tuvalu, the Philippines and Vietnam (Delpero, 2016). Notwithstanding the negligible contribution of their citizens to climate change, these countries are left with the crippling burden of adapting to its consequences.

Learning about climate change

Global climate change is a complex issue. Teaching climate change education is conceptually challenging (Shepardson et al., 2012). It requires teachers who are knowledgeable about climate change, its causes and consequences. Yet research indicates that educators have a limited understanding of climate change and climate action (Waldron et al., 2019). Developing awareness of and interest in climate change involves being attentive to the issue. Selby and Kagawa (2013) refer to 'a mindset of alertness and mindfulness' to climate change. While the issue is omnipresent, it is not visible on a day-to-day basis for those whose lifestyles remain unaffected, therefore easy to ignore. However, its visibility is stark for those affected by drought, for climate refugees and for those affected by extreme flooding. Many misconceptions about climate change from political sources, business interests and consumers in affluent societies need to be challenged. Teaching about climate change involves scientific knowledge, familiarity with appropriate pedagogy and resources, and confidence to explore the complexities of the subject, including moral and ethical aspects (Hestness et al., 2011). Several excellent climate education resources for primary teachers are listed at the end of this chapter.

Learning about climate change involves learning about its causes and consequences (Selby and Kagawa, 2013). Children should be able to understand the processes which contribute to climate change and discuss its impact on living things, people and the environment. An understanding of mitigation and adaptation strategies is also important. The mitigation dimension of climate change education addresses understanding the root causes of climate change. It involves the development of knowledge, skills and dispositions required by individuals and society to address these causes. Mitigation strategies in climate change education include education about: renewable energy; the design of ecotechnologies; energy conservation; impact of consumer patterns; value systems; and ideologies which have resulted in the emission of excessive greenhouse gases.

Adaptation seeks to lower the risks posed by the consequences of climate change. Adaptation involves learning to live with changing temperatures and seasons, extraordinary weather conditions, higher sea levels, more flooding and drought. Humans have always adapted to local climatic conditions including adaptations to crop types, building practices and cultural events. However, climate shifts including temperature, storm frequency and flooding may place unbearable pressure on communities. Those least responsible for climate change have few options available to them for adaptation. Adaptation measures include large-scale infrastructure changes and flood relief schemes as well as behavioural shifts such as water conservation and building of passive houses. Both climate change mitigation and adaptation will be necessary because even if emissions are dramatically decreased in the next decade, adaptation will still be needed to deal with the global changes that have already been set in motion (Selby and Kagawa, 2013).

Pedagogical approaches

While climate change is a geographical and scientific topic in its own right, it requires a holistic cross-curricular approach in classrooms and schools. Climate change education requires that teachers and children address a number of possible scenarios for our future.

This involves rethinking our current value systems and addressing common taken-forgranted assumptions about our world (Kagawa and Selby, 2010). Locally based enquiries help children to appreciate the relevance of climate change. For instance, children might investigate causes of local flooding and make connections with flooding scenarios elsewhere. Children need to be able to think critically and creatively about approaches to climate change mitigation and adaptation. They need to develop competencies which are transferable to current and future, certain and uncertain, situations (Wals, 2011).

Educating for hope

Issues such as climate change may potentially generate feelings of hopelessness and fear. As a topic it can be overladen with a sense of doom and catastrophe (Weintrobe, 2013). An eco-playful pedagogy (Witt and Clarke, 2014) takes a more hopeful perspective. Climate change education, handled in an age-appropriate and sensitive way (Sobel, 2008), equips and empowers children for the future. Such outcomes offer hope in uncertain times (Hicks, 2014). Climate change education should focus attention on futures and possible pathways to a sustainable future to promote hope in children (Ojala, 2015). Encouraging and inspiring individuals to take personal actions to mitigate climate change is encouraged by many (Lorenzoni et al., 2007; O'Neill and Nicholson-Cole, 2009; Wolf and Moser, 2011). Projects which facilitate children's actions are both hopeful and future-oriented. They encourage children to think positively about their futures. Children's agency can also be promoted through well-designed action projects. Some argue that individual actions are insufficient as climate change is a systemic problem requiring a complete overhaul of national and international values. While they may produce a feel-good factor, individual actions may not equip young people with skills for political involvement and collective climate change action (Schild, 2016). However, Waldron et al. (2019: 907) present a dynamic picture of climate change education which 'conceptualises children as present citizens capable of collective action'.

Climate change is considered by some teachers a controversial topic especially as it challenges the neo-liberal consumerism promoted by so many sectors of society. Yet if children are going to learn to think critically, teachers should be prepared to contest the prevailing dominant economic, political and social orthodoxies. Children need to have an opportunity to consider a different future and to imagine the world differently (Andreotti, 2006; Hicks, 2014).

Hicks (2014: 28–29) recommends a simple model for learning about global and local issues based on four dimensions: *knowing, feeling, choosing* and *acting* or in other words *the cognitive domain, the affective domain, decision-making* and *a sense of agency*. The four basic questions for children are: What do I need to know? What do I feel about this? What choices do I want to make? What can I do myself and with others?

1. Knowing

What do we think we know/need to know about climate change?
What are the main causes of climate change?
What are likely to be some of the consequences of climate change?

2. Feeling

What do I/we feel about climate change?
What are the concerns that we wish to share?
What are the hopes we might have?

3. Choosing

What are the options that appear to be facing us?
What do I/we want to see happening?
What should this school choose to work towards?

4. Acting

What do I/we therefore need to do?
What are others doing: school/home/community/elsewhere?
Who is able to support us in what we want to do?

The following case studies illustrate the use of this model in different ways. They provide examples of a 'critical, open-ended, holistic approach to climate change education' providing multiple spaces for reflection and engaging children with models of citizenship which embrace political action (Waldron et al., 2019: 13). These case studies illustrate the centrality of thinking geographically, enquiry-based learning and children's agency to conduct change.

Case study 5.2 Scoil Íde climate change

Scoil Íde, Corbally, Limerick adopts a whole-school approach to teaching climate change. Teachers acknowledge the requirement to teach atmosphere, climate and weather as part of the geography curriculum, but they do this in a cross-curricular manner. The school's work on weather and climate involves the children working as geographers, scientists, historians, scientists and mathematicians. Children use data to highlight how the climate is changing and the impact this is having on the natural world. As illustrated earlier in this chapter, children from this school conduct extensive work on seasonal changes, a key indicator of climate trends. The school's outdoor classroom and garden allow children to engage with nature in a real-life way, by growing vegetables, observing weather patterns and developing an appreciation for the natural environment. This allows children to experience the interconnected relationship between weather, seasons, nature and growth of vegetables.

Measuring the weather is serious business in Scoil Íde, Corbally. The children collect information through the weather station situated in the outdoor classroom. The weather station collects information about wind direction, rainfall, temperature and atmospheric pressure (Figure 5.9). It is powered by a solar panel on the front and sends all information wirelessly to a computer inside the school. Children (11–13 years) use this data to plot graphs and analyse weather patterns during the year, especially during the extreme winter months (Figure 5.10).

The school has invested in data loggers, or sensors. Children (aged 11–13) log natural light at the same time every morning, as well as rainfall, soil temperature and air pressure. They explore the relationship between this data on weekly weather and long-term climate change trends. The school uses Skylink-Pro, an online platform for reviewing, managing and reporting weather data. Its software gives real-time access to data collected by the school's weather station. It allows the school to measure, compare and share information across multiple sites. Data can also be stored and downloaded for historical analysis. Data from the weather station can be seen displayed on the Scoil Íde website and on its Twitter and Facebook accounts. When Hurricane Ophelia arrived in Ireland, the school's weather station recorded data and alerts were sent out on its Twitter feed.

Teaching powerful geography through topics 131

Figure 5.9 Picture of outdoor weather station in Scoil Íde, Corbally, Limerick

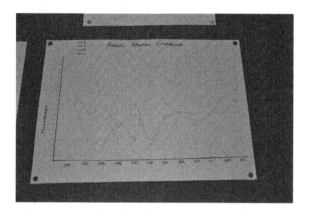

Figure 5.10 Long-term weather trends analysed by the children

Case study 5.3 The climate action project

St Augustine's NS participated in a Global Schools' Climate Action Project (www.climate-action.info). The overall aim of the initiative is to make children aware of the importance of taking action for climate change. Over the course of 4 weeks, 25 schools in 69 countries collaborated on climate change topics. Children from across the world conducted research, discussed their ideas and presented their findings via video and online presentations. During the final week of the project the children had an opportunity to share their work with other classes via Skype and to make meaningful connections.

This is a cross-curricular initiative exploring the geographical theme of climate change from a number of global vantage points. Children work as journalists and research the specific circumstances of climate change in their own regions and country. This includes causes, impacts and proposed actions. Their classrooms become newsrooms as the children prepare to broadcast their work and in this instance, they have a guaranteed global audience. This project is both child-centred and child-led as children drive the learning through questioning, research, discussion and presentation. Teachers act as guides and facilitators ensuring that sources are verified. Children complete their research tasks within agreed time frames.

Children in each country research climate change from their own local and national perspective, in light of their own interests and their own local circumstances. Hence the projects, discussions and outcomes are all different. This child-centred project is about making global connections. The project generates different responses and outcomes such as songs, dances, Lego creations, Minecraft, stop-motion videos and interviews.

Over the four-week period, children from St Augustine's NS made three informative and educational videos about their experiences of climate change.

In week 1 the children worked in teams to conduct some preliminary research on climate change and climate action. This was an exploratory week driven by the thoughts and ideas of the children. OneNote was used to collate and share their work.

Week 2 involved an exploration about how climate change affects the children in Galway and Ireland guided by the following enquiry questions:

1   What are the causes and effects of climate change in Galway/Ireland?
2   What natural disasters have recently happened in Ireland that have been caused by climate change?
3   Have there been more severe weather events in the last century in Ireland/Galway?
4   Are there any initiatives in Ireland to help stop climate change?

Week 3 was about actions for addressing climate change. Ironically the school had to close because of a Red Weather Warning issued by Met Éireann in anticipation of Hurricane Ophelia. The National Emergency Coordination Group met and gave advice to people to remain in their houses during the storm. Unfortunately, climate change means that scientists predict more of these weather events will affect Ireland in the future. The storm provided current and topical material for the children's research and Hurricane Ophelia featured prominently in the third video.

Week 4 of the Climate Action Project consisted of preparation for and participation in Skype calls with partner schools in New Jersey and Dubai (Figure 5.11). The children prepared presentations based on their research conducted over the previous three weeks.

The following is the children's record of the Skype calls, which has been posted on the project's website:

> On Tuesday 24th October we Skyped New Jersey and the class of Michael Dunlea. Michael teaches 7- and 8-year-olds in Tabernacle Elementary School in New Jersey. The children are learning Spanish. New Jersey was badly affected by Hurricane Sandy and 4,000 houses were destroyed. In New Jersey after Sandy and because of the flood risk to houses people have to raise their houses by 15 feet or else they won't get insured. The elderly and disabled people if they have a house like this they use an elevator to get up and down; it costs lots of money to make the house higher and a lot to install the elevator.
>
> They are studying climate change like our school. The school asked us some questions about what we are doing with the climate change project. After that we asked them some questions and what they do in their school. Their school is one of the best schools in their town because they are mindful of their water waste and reducing the waste of plastic. It was 2.19pm in Clontuskert and 9.19am in New Jersey when we spoke to Mr Dunlea's class. That is a five-hour time difference. START is Stafford

Teachers and Residents Together and this group was founded by Mr Dunlea in response to the devastation of Hurricane Sandy.

On Wednesday 25th October we Skyped into GEMS School in Dubai with Mrs Shafaque Riaz. GEMS is also a Microsoft Showcase School. People in Dubai speak Arabic. Many of the parents work in the airport which is near GEMS School. They have more they 4,000 children in their school. They have 105 different nationalities. During our call to Dubai it was 37 degrees. They have a lot of traffic every day in Dubai.

When the children were researching for the Climate Action Project they found that they have had more sandstorms this summer and they predict they are going to have more rain in winter. They have a Minister for Climate Change in Dubai like we do in Ireland. They reuse water from the buses after they wash them. School starts at 7.30 in the morning and ends at 4.00pm. They are testing flying cars and self-driving cars in Dubai. They are trying to put the right thing into the right bins. Every month they go to the beach to clean it.

The children were not in the classroom at the time so we did not get to meet them. Mrs Riaz showed us some photos of Dubai and she asked us questions about Ireland. We told her about Hurricane Ophelia and what we have been doing in Clontuskert to help stop climate change. We told her about the bins and about the green school and what we do. There national sport is football. We spoke to Mrs Riaz for 40 minutes.

The project is based on a number of important principles.

It is cross-curricular.
It embraces 21st-century learning skills such as knowledge-building, use of ICT, problem-solving and self-monitoring.
It involves the use of technology including computers, tablets and Web 2.0 tools.
Collaborative learning within and between schools is essential.
It is framed by the Sustainable Development Goals, which are discussed in greater detail in Chapter 8.

The children's work is available to view on the project's web site: www.climate-action.info.

Reflections from children

1   What did I learn?
    - *I learned how much climate change is affecting our world, and that animals are becoming extinct and humans are affected by it.*
    - *I learned that the burning of fossil fuels would damage our atmosphere, reducing oxygen and could kill us. If we destroy all the trees we will have bad air.*
    - *I learned that all our rubbish is killing so many different animals and our climate in Ireland was once great but because of our burning of fossil fuels we are destroying our Earth at an accelerated pace.*
    - *I learned that the world is getting warmer every year because of climate change.*
    - *I learned about climate change, what it is and how to try to stop it.*
    - *I learned that all of the gases that we put out into the air are destroying the planet.*
    - *I learned that climate change is a life-changing situation with big storms that can destroy everything so we have to prevent this from happening.*

134 *Teaching powerful geography through topics*

Figure 5.11 Learning about the impacts of climate change in Dubai

2   What actions should our school take in light of our climate change project?
    - *We should put leftover food items in the compost bin.*
    - *I think we should carpool even more than we already do and think about how we could reuse our rubbish, like with yogurt pots and paper we might be able to do art with them.*
    - *We should encourage everyone to carpool, not just in school but in workplaces, so not as many gases are going into the air and get the bus or some form of public transport.*
    - *Our school should continue to get the word about our project out to the world, so people can see how we are damaging it.*
    - *We could explain how to reduce, reuse and recycle to children when they are young so they know and understand the bin system.*
    - *We should use more electric cars.*

3   How will my behaviour change as a result of this project?
    • *I will look after trees and I will plant more trees.*
    • *I will be more careful with things that can harm the environment and the atmosphere.*
    • *This project has influenced me to help me to care more for the environment and to prevent climate change.*
    • *I will be more careful with recycling and reusing plastic because it might help our planet.*
    • *We understand now what's bad and what's good for the environment so this will help us to be more mindful of how to care for our planet in the future.*
    • *I think I could reduce my waste items and think about how they could come in handy and bake more things rather than buying items in packets that have to be thrown away.*
4   How would you advise the Minister responsible for Climate Action and Environment on actions that the government should take?
    • *I would say to make a bigger effort to encourage electric cars and to stop items being wrapped in so many layers of plastic material.*
    • *It is also important for the government to create more awareness of how our Earth is being slowing destroyed by fossil fuels and how animals are being killed. This is a big concern and needs to be dealt with.*
    • *The government should ban the use of things that are harmful to the atmosphere like cigarettes, fossil fuels, etc.*
    • *The Minister should look more into climate action and reduce the price of electric cars to encourage more people to buy them.*
    • *The Minister should increase the bin prices so more people will recycle and stop dumping. The government should also put more public bins with recycling and landfill side by side, so people can separate their waste, in big towns like Ballinasloe.*

These reflections demonstrate evidence of powerful geographical knowledge emerging through knowledge building, critical thinking, real world problem-solving, collaboration, team work and global competency as illustrated in Table 5.7. Through collaborative learning, children began the process of making links between unusual weather conditions and climate change. This initiated a discussion on action. The next case study illustrates action projects addressing climate change.

Case study 5.4 Lego robotics

Lego has for generations provided opportunities for hands-on, kinaesthetic learning that promotes collaboration and team-building skills among children. Working with Lego complements many curricular areas including geography in a fun and engaging way. The introduction of Lego robotics has taken the use of Lego to an entirely new level. Lego robotics introduces children to technology and the opportunities for exploring cross-disciplinary concepts are numerous (Scaradozzi et al., 2015). Robots fascinate children and are common currency on TV in movies and literature. Their application in modern society in areas such as medicine, exploration and science are well covered by the media.

The FIRST® LEGO® League (FLL) is an international competition involving children and teenagers aged 11 to 16 from 80 countries. Every year, FIRST LEGO League releases a challenge, which is based on a real-world topic. Previous challenges have been based on geographical topics such as climate, transportation and water. One year the challenge was related to natural disasters. Children were invited to learn about a natural event which could cause a natural disaster in a community. They were invited to consider the following questions:

How could this force of nature cause a natural disaster in your chosen community?
What problems would it cause? Would it harm people, property or both?
Is the community continuously at risk or during certain times of the year? Is the risk predictable?
How do people prepare?
How are people warned?
Who provides emergency food, water and shelter? How?
Who clears the debris and rebuilds? How long does it take?

The children are then asked to design an innovative solution that helps people prepare, stay safe or rebuild, and present their solution at a local Lego tournament.

The children from Clontuskert NS chose the problem of flooding as their local town Ballinasloe was badly affected in recent floods. This is their report:

> The recent flooding in Ballinasloe was a natural disaster that we felt strongly about. It had a devasting effect on the people whose houses and businesses were ruined. Local businesses such as John Murray's The Bread Basket and journalists such as Ken Kelly gave a graphic account of the floodwater's path through the town and the aftermath. We rang Birdwatch Banagher to find out what kind of effect it has on the animal population. She gave us some very interesting facts that we added to our project. We feel it is very important to share our work with the public and for the last three years we have attended the Galway Science Festival.
>
> We really want to educate the public about the danger of flooding and ways to prevent it. We made a film using photographs taken by a local photographer, Robert Ridell. We invited visitors to our stand to take our survey to see what they knew about floods and if they had any ideas about a robot we could build to stop flooding.
>
> - 86% did not know they could drown in 6 inches/150mm of water.
> - 33% did not know how to prevent flooding.
> - 29% have been affected by the flood.
> - 5% lost everything in their house.
>
> All of our research guided us towards the idea of making an app. Our idea is to update the public about hazards on the road or in their local area. We rang Galway County Council to find out what kind of notification system they had in place to notify the public about these events. They had nothing in production yet but were very interested in helping us with our work. They said that Twitter and Facebook are the main ways for notifying the public. They also have areas mapped out in Galway that are at risk of flooding. Our app will be especially useful if you are travelling. You can put in your starting destination point and then it would notify you if the route you were taking is blocked and why it is blocked if there was a fallen tree, if there were floods or if there was an electric wire down. We plan on

linking with Waterways Ireland as they have a camera system in development that can measure water levels in some rivers such as the Suir in Co. Tipperary. As you know the weather at the moment is very unpredictable so this app would be extremely useful. We have experimented with app software such as MIT app inventor. The app maker is a site that is fun and exciting to use. We are very confident that in the future our app will be in production. We have also applied for a grant from the county council.

Lego robotics promotes hands-on meaningful learning as children engage with concepts in a concrete way. Working with robotics embraces constructivist approaches to learning (Butler et al., 2015). Teachers from St Augustine's NS use robotics and coding as tools to embrace curricular engagement and to ensure children cultivate lifelong skills. The Lego robotics system WeDo™ allows children to build and create their own model robot. Computer programming designed by the children manages the robot's movement. From a geographical perspective Lego construction material enables children to come to know their surrounding world by recreating it. Both the FIRST LEGO League and FIRST® LEGO® League Jr promote core values such as group work, building self-confidence, knowledge creation, teamwork and life skills. Working with robotics demonstrates how a traditional curriculum can be delivered from a 21st-century perspective.

The Lego kits produced by the education division of the Danish toymaker include sensors and motors that respond to commands remotely through simple coding instructions. In St Augustine's NS the children use the WeDo™ Lego robotics system, which is a set of pieces and mechanical parts used to build and design Lego models. The set contains robot bricks, two sensors, Lego USB hub and a motor. The set comes with easy-to-use icon-based software providing an intuitive programming environment with building instructions, programming examples and activity tips. Initially the children are taught a basic working knowledge of robotics in an age-appropriate way. They then work in teams to design, construct and program a mobile robot.

FIRST LEGO League Jr is an exciting Science, Technology, Engineering and Maths (STEM) programme for 6–9-year-olds. FIRST LEGO League Jr is designed to introduce STEM concepts to children aged 6 to 10 while exciting them through the Lego brand they know and love. FIRST is an acronym for 'For Inspiration and Recognition of Science and Technology'. Each year a theme relevant to the world around them is chosen as a focus for the children's work. The teams research the topic, displaying their ideas on a Show Me poster and they also build a motorised Lego model. FIRST LEGO League Jr rapidly develops teamwork, design, programming and communication skills, but more importantly it is great fun and makes the children really enjoy learning.

For FIRST LEGO League Jr (Disaster Blaster) the children learnt about volcanoes, floods and snow storms and came up with solutions to problems caused by these disasters. They designed and constructed a TEAM MODEL using Lego bricks and motorised moving parts from the Lego Education WeDo™ 2.0 kits. They designed various tasks which involved moving people to a safe area. Certain obstacles including trees and fallen wires had to be circumnavigated by the robot. Houses were built on stilts and an ambulance battled to get people to safety. Small animals were moved to a safe area. Figure 5.12 shows pictures from the FISRT LEGO League Jr HYDRO DYNAMICS challenge, which is all about water – how we find, transport, use or dispose of it.

138 *Teaching powerful geography through topics*

Figure 5.12 Exploring the dynamics of water through Lego

Case study 5.5 Building a flood defence

Educating for hope includes problem-posing and problem-solving whereby children collaborate, investigate and work as problem-solvers. As part of a climate change adaptation process, children from Mary, Help of Christians, Girls' National School, Navan Road, Dublin investigated the advantages and disadvantages of different materials which could be used in flood defences. They tested a number of materials including: Lego, modelling clay, sponges, cork, lollipop sticks and sand (Figure 5.13). Each group designed and built a flood wall based on the nature of their material with the aim of keeping water out for the longest period possible. Each wall was placed along a line on a baking tray. The defence which kept the 'teddy dry' the longest was considered the most effective. From a scientific perspective a fair test was guaranteed by ensuring all baking trays were the same size, all defence lines were at the same distance and the amount of water poured into each test was the same and poured at the same time. The children discovered that modelling clay was the most effective followed by the sandbags and Lego. The least effective material for defence was the sponge material, followed by the cork and lollipop sticks.

Figure 5.13 Building a flood defence

What is so powerful about weather and climate change education?

Weather and climate are examples of powerful primary geographical topics. They are not luxury topics to be pushed to the margins of the curriculum. It is important to remember that climate change is one of the greatest challenges facing the world and its citizens today. Yet there are widespread misconceptions within every sector of society including agriculture, media and politics.

Education in general and primary geography in particular have powerful roles to play. As human activity is one of the main causes of climate change everyone can and should take action against global warming. While we share the responsibility for greenhouse emissions we are also part of the solution. Nevertheless, it is important not to burden children with messages of doom and gloom. More importantly children need to see that they have the power to make informed choices about their behaviour in line with their conceptions of possible, probable and preferred futures (Hicks, 2017).

Teacher understanding of these geographical concepts is crucial. If teachers are critically thoughtful about weather and climate justice education, they can help children to become critically thoughtful. Once children and teachers are well informed they can consider how to behave and act as a result of their conceptual understanding. It is the study of primary geography which provides this 'powerful knowledge'. However, it can only be powerful if it informs behaviours and attitudes. Changing behaviour in an informed manner is the important aspiration of climate change education. Using the framework illustrated in Chapter 1, Table 5.7 demonstrates how teaching weather and climate change constitutes powerful geographical knowledge.

Table 5.7 Teaching weather and climate change as powerful geographical knowledge

Maude's five types of powerful knowledge	21st-century competencies	Teaching weather and climate change powerfully
Knowledge that provides children with new ways of looking at the world	Knowledge building	Thinking geographically and developing conceptual knowledge about weather and climate
Knowledge that provides children with powerful ways to analyse, explain and understand the world	Critical thinking	Asking geographical questions (making connections, understanding interconnections and developing well-informed understanding about weather and climate change)
Knowledge that gives children some power over their own knowledge	Real-world problem-solving	Formulating geographical solutions to climate change
Knowledge that enables young people to follow and participate in debates on significant local, national and global issues	Collaboration and teamwork	Working together to extend geographical understanding about weather and climate change though collaboration, team work, action-based projects and sharing new understanding
Knowledge of the world	Global competency	Becoming an ambassador for weather and climate change/justice education
		Sharing findings from classroom research on climate issues with a broader audience

Climate justice education is ultimately about changing behaviour; thus, it has the potential to be transformative. The climate change newspaper supplement created by children from St Brendan's NS, Eyrecourt discussed in Chapter 2 is an example of children's work going beyond the classroom walls to inform the local community. Through this work children demonstrated global competency in their effectiveness in becoming ambassadors for climate justice education. *The Book of Climate Bells* described in Chapter 7 represents a symbolic call to action by all citizens.

Conclusion

Reasons for teaching weather and climate are compelling. Climate change is one of the dominant challenges of our time. Climate justice education is absolutely essential for 21-century citizens and thinkers. It provides opportunities for rethinking the world, for futures education and for encouraging 'out-of-the-box' thinking (Glasser, 2007). Learning is essential for the resolution of climate change issues. Children need to be considered citizens capable of collective action. The goal for climate change education as articulated by Kagawa and Selby (2010: 4) is that 'the learning environment can be seized to think about what really and profoundly matters, to collectively envision a better future, and then to become practical visionaries in realizing that future'. Understanding climate change has to begin with an understanding of weather and climate. Case studies in this chapter showcase some amazing work whereby children are learning to research, to think creatively and to share their innovative practice with a broader public.

Exercise 5.1 Personal reflection for teachers

Why is it important to teach about weather and climate change?
How does your school teach weather and climate change?
Do you engage with the weather as it occurs outside (through experience, measurement, use of instruments made by the children?)
Do you use unusual weather events as opportunities for teaching geography?
What resources and/or CPD do your staff need to teach topics related to weather and climate change?

Further resources

Bowles, R. (2010) Weather and climate. In Scoffham, S., ed., *Primary geography handbook*. Sheffield: Geographical Association, 230–245.
Collis, S. (2017) *Super schemes: Climate and biomes*. GA: Sheffield.
Danks, F. and Schofield, J. (2013) *The wild weather book: Loads of things to do outdoors in rain, wind and snow*. London: Frances Lincoln.
Donnelly, J. (2018) *The great Irish weather book*. Dublin: Gill Books.
Dunlop, S. (2017) *Weather: a very short introduction*. Oxford: OUP.
Maslin, M. (2014) *Climate change: A very short introduction*. Oxford: OUP.

Websites

Met Éireann is Ireland's National Meteorological Service: www.met.ie/education/.

MetLink is the educational website of the Royal Meteorological Society. It is a site for teachers and children, providing teaching materials and resources for teaching weather and climate to children. It has lesson plan downloads for weather and climate, information on the latest Intergovernmental Panel on Climate Change (IPCC) findings, and much more: www.metlink.org/.

Creating Futures: This climate change education resource for senior primary classrooms consists of ten lessons to inspire enquiry, creativity and cooperation. Resources are cross-curricular and include lesson plans, activities, worksheets, photographs, web links and lots more. The first three lessons examine the science of climate change exploring a historical perspective on the greenhouse effect and evaluating the evidence relating to the causes and consequences of climate change. Lessons 4–6 explore the impact of climate change on people, particularly those in developing countries and on biodiversity. The last three look at what we can do about climate change including designing more sustainable transport, making tough decisions and showing leadership. The final lesson is a writing activity which works as an assessment. The lessons have been designed to fulfil aspects of the science, history, geography, SPHE, English, drama, art and maths curriculum: www.dcu.ie/chrce/news/2016/sep/creating-futures-climate-change-education-for-senior-primary.shtml.

Action Aid: Power Down: A climate change toolkit for primary schools. This is designed for senior classes. It includes a teacher's booklet with classroom ideas and photographs showing the impact and consequences of climate change. On the reverse side of each photocard are key facts, discussion, questions and top tips: www.globalfootprints.org/files/zones/act/schools_powerdown_primary.pdf.

Trócaire: www.trocaire.org/education/climate-change/primary.

Oxfam education resources on climate change provide climate change educators with case studies of various programmes being implemented in developing countries, together with related lesson plans and activities: www.oxfam.org.uk/education/resources/category.htm?20.

Climate 4 Classrooms is a resource developed by the Royal Geographical Society in collaboration with climate scientists and using data from the latest research, including the Intergovernmental Panel on Climate Change (IPCC). It offers teaching and learning resources that enable children to learn about the science of climate change, investigate possible global and national futures and explore global and local solutions. Each module has clear learning outcomes, activity plans and child activity sheets: http://uk.climate-4classrooms.org/teaching-resources.

Climate Change in the Classroom: An excellent CPD resource for teachers (Selby and Kagawa, 2013): http://unesdoc.unesco.org/images/0021/002197/219752e.pdf.

Eco Detectives: This resource provides environmental and climate change investigations for primary schools. It is produced by the Department of the Environment, Heritage and Local Government along with the Centre for Human Rights and Citizenship Education, Dublin City University (DCU). The resource pack contains teacher books, with interactive activities and resource material, including guides for children, photo cards and an Eco Detective game: www.askaboutireland.ie/learning-zone/primary-students/infants/environment/teachers-resources-eco-de/.

Online videos for teachers

Before the Flood video by Leonardo De Caprio: https://archive.org/details/youtube-9OCkXVF-Q8M.

Drop in the Ocean? Ireland and Climate Change video: www.trocaire.org/drop.

Articles published in the Geographical Association's journal *Primary Geography*

Aspin, V. (2018) A week of rain. *Primary Geography*, 96, 24–25.
Campbell, L. (2018) Reading the weather. *Primary Geography*, 96, 10–11.
Fordham, R. (2011) 2010: An extraordinary year for global weather. *Primary Geography*, 76, 14–15.
Green, A. (2018) Energy matters. *Primary Geography*, 96, 20–21.
Greenwood, H. (2018) A climate change assembly. *Primary Geography*, 96, 22–23.
Hatwood, R. (2018) Reacting to the weather. *Primary Geography*, 96, 14.
Knight, S. (2011) Tornadoes in your classroom. *Primary Geography*, 76, 16–17.
MacNaughton, A. (2015) Tornadoes in your classroom. *Primary Geography*, 88, 19–21.
McLellan, G. (2018) Writing the flood. *Primary Geography*, 96, 8–9.
Owen, C., Witts, S. and Burnett, L. (2018) Can we play outside today? *Primary Geography*, 96, 18–19.
Owens, P. (2018) Weather glossary. *Primary Geography*, 96, 27–31.
Scoffham, S. (2018) Extreme weather. *Primary Geography*, 96, 6–7.
Spear, P. (2018) Get in the picture about climate change. *Primary Geography*, 96, 26–27.
Thorpe, J. (2018) Weather and climate in the curriculum. *Primary Geography*, 96, 12–13.
Whitehouse, A. (2008) Weather around the world. *Primary Geography*, 65, 37–38.
Whittle, J. (2018b) Reading the weather. *Primary Geography*, 96, 5.

Children's literature

Children's literature provides a myriad of opportunities for teaching geographical concepts and vocabulary (Whittle, 2018). Here I include examples of children's fiction and non-fiction which deal with the geographical topics of weather.

Children's fiction and non-fiction literature about the weather

Fiction picture books about weather

Oh Say Can You Say What's the Weather Today? (2011) Tish Rabe and Aristides Ruiz, Harper Collins Children's Books.
Hello World Weather (2016) Jill McDonald, RH Children's Books.
Maisy's Wonderful Weather Book (2011) Lucy Cousins, Walker.
Alfie: Alfie Weather (2002) Shirley Hughes, Red Fox.
My Friend the Weather Monster (2012) Steve Smallman and Bruno Merz (illus), QED Publishing.
If Frogs Made Weather (2005) Marion Dane Bauer and Dorothy Donohue (illus), Holiday House.
Marvelous Cornelius: Hurricane Katrina and the Spirit of New Orleans (2015) Phil Bilder and John Parra, Chronicle Books.
After the Storm (2011) Nick Butterworth, Harper Collins Children's Books.
Flood (2014) A.F. Villa, Curious Fox.
Hurricane (1992) David Wiesner, Houghton Mifflin.
Cloudland (1999) John Burningham, Red Fox.
Cloudy with a Chance of Meatballs (2012) Judi Barrett, Little Simon.
The Cloudspotter (2016) Tom McLaughlin, Bloomsbury Children's Books.
Little Cloud (2001) Eric Carle, Puffin Books.
Clouds (2008) Anne Rockwell, Harper Collins.
Lila and the Secret of Rain (2009) David Conway and Jude Daly (illus), Frances Lincoln Children's Books.
Rainy Day (2000) Emma Haughton and Angelo Rinaldi (illus), Doubleday Children's Books.
The Rainy Day (2012) Anna Milbourne and Elena Temporin, Usborne Publishing Ltd.
Splish, Splash Splosh: A Book about Water (2014) Mick Manning and Brita Granström, Franklin Watts.

The Little Raindrop (2014) Joanna Gray and Dubravka Kolanovic (illus), Sky Pony Press.
The Drop Goes Plop: A First Look at the Water Cycle (1998) Sam Godwin and Simone Abel (illus), Wayland.
Rain (Whatever the Weather) (2014) Carol Thompson, Child's Play (International) Ltd.
Monsoon (2003) Ulma Krishnaswami and Jamel Akib (illus), Farrar Straus Giroux.
Storm in the Night (1996) Mary Stolz and Pat Cummings (illus), Harper & Row Ltd; 1st Harper Trophy Ed edition.
Who Likes the Rain? (2007) Etta Kaner and Marie Lafrance (illus), Kids Can Press.
Rain (2016) Sam Usher, Templar Publishing.
Who Likes the Sun? (2007) Etta Kaner and Marie Lafrance (illus), Kids Can Press.
Sun (Whatever the Weather) (2014) Carol Thompson, Child's Play (International) Ltd.
Sun (2017) Sam Usher, Templar Publishing.
Snow (2014) Sam Usher, Templar Publishing.
Snow (Whatever the Weather) (2014) Carol Thompson, Child's Play (International) Ltd.
The Ice Bear's Cave (2014) Mark Haddon and David Axtell, Harper Collins Children's Books.
Snowflake Bentley (2009) Jacqueline Briggs, Martin and Mary Azarian (illus), Houghton Mifflin Harcourt.
The Windy Day (2012) Anna Milbourne and Elena Temporin, Usborne Publishing Ltd.
Feel the Wind (2001) Arthur Dorros, William Morrow.
Rosie's Hat (2015) Julia Donaldson and Anna Currey (illus), Pan Macmillan Publishers Ltd.
Like a Windy Day (2008) Frank Asch and Devin Asch (illus), Houghton Mifflin.
The Sunny Day (2012) Anna Milbourne and Elena Temporin, Usborne Publishing Ltd.

Non-fiction children's literature about the weather

A Journey through the Weather (2016) Steve Parker and John Haslem, QEB Publishing.
Weather in 30 Seconds (2016) Jen Green and Tom Wooley, Ivy Kids.
Cloud Dance (2003) Thomas Locker, Turtleback Books.
A Drop in the Ocean: The Story of Water (Science Works) Jacqui Bailey and Mathew Lilly (illus), A & C Black Publishers Ltd.
How the Weather Works (2011) Christiane Dorion and Beverley Young, Templar.
Under the Weather: Stories about Climate Change (2012) Tony Bradman, ed., Frances Lincoln.

Children's fiction and non-fiction literature about climate change

Winston of Churchill: One Bear's Battle Against Global Warming (2010) Jean Davies Okimoto and Jeremiah Trammell (illus), J. Scholastic Canada, Limited.
Perry the Polar Bear Goes Green: A Story about Global Warming (2010) Olive O'Brien and Nina Finn-Kelcey (illus), BPR Publishers.
Perry the Playful Polar Bear (2009) Olive O'Brien and Nina Finn-Kelcey (illus), BPR Publishers.
Little Polar Bear (2011) Hans de Beer, North South Books.
The Polar Bear's Home (2008) Lara Bergen, Little Simon.
Why are the Ice Caps Melting? The Dangers of Global Warming (2007) Anne Rockwell and Paul Meiser (illus), Harper Collins.
Hot Air (2013) Sandrine Dumas Roy and Emmanualla Houssais, Phoenix Yard Books.
The Problem of the Hot World (2015) Pam Bonsper and Dick Rink (illus), (Kindle).
The Magic School Bus and the Climate Challenge (2014) Joanna Cole and Bruce Degen (illus), Scholastic Press.
Who Turned Up the Heat? Eco Pig Explains Global Warming (2009) Lisa S. French and Barry Gott (illus), Looking Glass Library.
Please Don't Paint Our Planet Pink: A Story for Children and Their Adults (2014) Gregg Kleiner and Laurel Thompson (illus), Cloudburst Creative.
The Tantrum that Saved the World (2018) Megan Herbert and Michael E. Mann, World Saving Books.

Non-fiction for older children

Our Choice: How We Can Solve the Climate Crisis (Young Reader Edition) (2009) Al Gore, Puffin Books.
Our Earth: How Kids are Saving the Planet (2010) Janet Wilson, Second Story Press.
It's Getting Hot in Here (2016) Bridget Heos, Houghton Mifflin Harcourt.
Global Warming (2013) Seymour, Simon Collins.
Eyes Wide Open: Going Behind the Environmental Headlines (2104) Paul Fleischman, Candlewick Press.
The Down to Earth Guide of Global Warming (2007) Laurie David and Cambria Gordon, Orchard Books.
Our House is Round: A Kid's Book about Why Protecting Our Earth Matters (2012) Yolanda Kondonassis and Joan Brush (illus), Sky Pony.
Basher Science: Climate Change (2015) Simon Basher, Kingfisher.

Autumn picture books

Animal Seasons: Squirrel's Autumn Search (2013) Anita Loughrey and Danial Howarth (illus), QED Publishing.
Autumn Is Here (2012) Heidi Pross Gray, CreateSpace Independent Publishing Platform.
Autumn (2015) Ailie Busby, Child's Play (International) Ltd.
Leaf Man (2005) Lois Ehlert, Harcourt.
Fletcher and the Falling Leaves (2008) Julia Rawlinson and Tiphanie Beeke (illus), Greenwillow Books.
What Can You See in Autumn? (2014) Sian Smith, Raintree (UK Edition).

Winter picture books

One Winter's Day (2013) M. Christina Butler and Tina Macnaughton (illus), Good Books.
The Shortest Day: Celebrating the Winter Solstice (2014) Wendy Pfeffer and Jesse Reisch (illlus), Puffin Books.
The Snowy Day (2011) Anna Milbourne and Elena Temporin, Usborne Publishing Ltd.
The Winter Fox (2017) Timothy Knapman and Rebecca Harry (illus), Nosy Crow.
What Can You See in Winter? (2014) Sian Smith, Heinemann.
William's Winter Wish (2012) Gillian Shields and Rosie Reeve (illus), Macmillan Children's Books.
Winter (2015) Ailie Busby, Child's Play (International) Ltd.
Winter Senses (2016) Dee Smith, CreateSpace Independent Publishing Platform.

Summer picture books

At the Beach (2007) Roland Harvey, Allen and Unwin.
Come Away from the Water, Shirley (2008) John Burningham, Red Fox.
Magic Beach (2006) Alison Lester, Allen and Unwin.
Mama Is It Summer Yet? (2018) Nikki McClure, Harry N. Abrams.
Mouse's First Summer (2013) Lauren Thompson and Buket Erdogan (illus), Little Simon.
Sneakers the Seaside Cat (2005) Margaret Wise Brown and Anne Mortimer (illus), Harper Collins.
Summer (2015) Ailie Busby, Child's Play (International) Ltd.
Summer Is Here (2013) Heidi Pross Gray, CreateSpace Independent Publishing Platform.

Spring picture books

Rabbit's Spring Adventure (2013) Anita Loughrey and Daniel Howarth (illus), QED Publishing.
Spring (2015) Ailie Busby, Child's Play (International) Ltd.
Spring Is Here (2013) Heidi Pross Gray, CreateSpace Independent Publishing Platform.
Spring Is Here: A Bear and Mole Story (2012) Will Hillenbrand, Holiday House.
What Can You See in Spring? (2014) Sian Smith, Heinemann.

Picture books about the seasons

Skip through the Seasons (2007) Stella Blackstone and Maria Carluccio (illus), Barefoot Books.
The Story Orchestra: Four Seasons in One Day (2016) Jessica Courtney-Tickle, Frances Lincoln Children's Books.
Goodbye Summer, Hello Autumn (2016) Kenard Pak, Henry Holt and Co.

References

Anderson, A. (2012) Climate change education for mitigation and adaptation. *Journal of Education for Sustainable Development*, 6(2), 191–206.
Andreotti, V. (2006) Soft versus critical global citizenship education. *Policy and Practice: A Development Education Review*, 3(4), 40–51.
Boon, H.J. (2016) Pre-service teachers and climate change: A stalemate? *Australian Journal of Teacher Education*, 41(4), 39–63.
Butler, D., Marshall, K. and Leahy, M. (Eds.) (2015) *Shaping the future: How technology can lead to educational transformation*. Dublin: Liffey Press.
Choi, S., Niyogi, D., Shepardson, D.P. and Charusombat, U. (2010) Do earth and environmental science textbooks promote middle and high school students' conceptual development about climate change? Textbooks' consideration of students' misconceptions. *Bulletin of the American Meteorological Society*, 91, 889–898.
Dawson, V. (2015) Western Australian high school students' understandings about the socioscientific issue of climate change. *International Journal of Science Education*, 37(7), 1024–1043.
Delpero, C. (2016) *New subject at school: Climate change. The road to Paris: Science for smart policy*. Paris: International Council for Science (ICSU).
Glasser, H. (2007) Minding the gap: The role of social learning in linking our stated desire for a more sustainable world to our everyday actions and policies. In Wals, A.E. ed., *Social learning towards a sustainable world*, Wageningen: Wageningen Academic Publishers, 35–62.
Gough, A. and Gough, N. (2016) The denaturation of environmental education: Exploring the role of ecotechnologies. *Australian Journal of Environmental Education*, 32 (SI1), 30–41.
Hatwood, R. (2018) Reacting to the weather. *Primary Geography*, 96, 14.
Hestness, E., Randy McGinnis, J., Riedinger, K. and Marbach-Ad, G. (2011) A study of teacher candidates' experiences investigating global climate change within an elementary science methods course. *Journal of Science Teacher Education*, 22(4), 351–369.
Hicks, D. (2014) *Educating for hope in troubled times: Climate change and the transition to a post-carbon future*. London: Trentham Books Limited.
Hicks, D. (2017) Zero-carbon Britain: looking to the future? *Teaching Geography*, 42(2), 69–71.
Kagawa, F. and Selby, D. (2010) Introduction. In Kagawa, F. and Selby, D., eds., *Education and climate change: Living and learning in interesting times*, London, Routledge, 1–11.
Krasny, M.E. and DuBois, B. (2019) Climate adaptation education: Embracing reality or abandoning environmental values. *Environmental Education Research*, 25(6), 883–894.
Lambert, J.L., Lindgren, J. and Bleicher, R. (2012) Assessing elementary science methods students' understanding about global climate change. *International Journal of Science Education*, 34(8), 1167–1187.
Lane, R. and Catling, S., 2016. Preservice primary teachers' depth and accuracy of knowledge of tropical cyclones. *Journal of Geography*, 115(5), 198–211.
Lorenzoni, I., Nicholson-Cole, S. and Whitmarsh, L. (2007) Barriers perceived to engaging with climate change among the UK public and their policy implications. *Global Environmental Change*, 17(3–4), 445–459.
Mochizuki, Y. and Bryan, A., 2015. Climate change education in the context of education for sustainable development: Rationale and principles. *Journal of Education for Sustainable Development*, 9(1), 4–26.

National Oceanic and Atmospheric Administration (NOAA) available on oceanservice.noaa.gov (accessed 12 January 2020).

NCCA (2013) Learning and developing through play www.ncca.biz/Aistear/pdfs/Guidelines_ENG/Play_ENG.pdf

Ojala, M. (2015) Hope in the face of climate change: Associations with environmental engagement and student perceptions of teachers' emotion communication style and future orientation. *The Journal of Environmental Education*, 46(3), 133–148.

O'Neill, S. and Nicholson-Cole, S. (2009) 'Fear won't do it': Promoting positive engagement with climate change through visual and iconic representations. *Science Communication*, 30(3), 355–379.

Papadimitriou, V. (2004) Prospective primary teachers' understanding of climate change, greenhouse effect, and ozone layer depletion. *Journal of Science Education and Technology*, 13(2), 299–307.

Rooney, T. (2016) Weather worlding: Learning with the elements in early childhood. *Environmental Education Research*, 24(1), 1–12.

Scaradozzi, D., Sorbi, L., Pedale, A., Valzano, M. and Vergine, C. (2015) Teaching robotics at the primary school: An innovative approach. *Procedia-Social and Behavioral Sciences*, 174, 3838–3846.

Schild, R. (2016) 'Environmental citizenship: What can political theory contribute to environmental education practice? *The Journal of Environmental Education* 47(1), 19–34. doi: 10.1080/00958964.2015.109241

Scoffham, S. (2018) Extreme weather. *Primary Geography*, 96, 6–7.

Shepardson, D.P., Niyogi, D., Roychoudhury, A. and Hirsch, A. (2012) Conceptualising climate change in the context of a climate system: Implications for climate and environmental education. *Environmental Education Research*, 18(3), 323–352.

Sobel, D. (2008) *Childhood and nature: Design principles for educators*. Portland, ME: Stenhouse Publishers.

Stevenson, R.B., Nicholls, J. and Whitehouse, H. (2017) What is climate change education? *Curriculum Perspectives*, 37(1), 67–71.

Tsaparlis, G. (2003) 'Chemical phenomena versus chemical reactions: Do students make the connection?' *Chemistry Education: Research and Practice*, 4(1), 31–43.

Trócaire (2017) Activist toolkit: Climate justice campaign www.trocaire.org/sites/default/files/pdfs/campaigns/climate-toolkit.pdf

Tutu, D. and Robinson, M. (2011) Climate change is a matter of justice. *The Guardian*, 5 December. www.guardian.co.uk/environment/2011/dec/05/climate-change-justice?intcmp=122.

UNESCO (2016) Education for people and planet: Creating sustainable futures for all. Global education monitoring report. Retrieved from http://gem-report-2016.unesco.org/en/home/

United Nations (2017) *Goal 13: Take urgent action to combat climate change and its impacts*. Retrieved from: www.un.org/sustainabledevelopment/climate-change-2

Waldron, F., Ruane, B., Oberman, R. and Morris, S. (2019) Geographical process or global injustice? Contrasting educational perspectives on climate change. *Environmental Education Research*, 25(6), 895–911.

Wals, A.E. (2011) Learning our way to sustainability. *Journal of Education for Sustainable Development*, 5(2), 177–186.

Weintrobe, S. ed., (2013) *Engaging with climate change: Psychoanalytic and interdisciplinary perspectives*. London: Routledge.

Whittle, J. (2018) Reading the weather. *Primary Geography*, 96, 5.

Witt, S. and Clarke, H. (2014) Seeking to unsettle student teachers' notions of curriculum: Making sense of imaginative encounters in the natural world. *Proceedings of Teacher Education for Equity and Sustainability (TEESNet) Seventh Annual Conference, Liverpool Hope University*, http://teesnet.ning.com/page/resources

Wolf, J. and Moser, S.C. (2011) Individual understandings, perceptions, and engagement with climate change: Insights from in-depth studies across the world. *Wiley Interdisciplinary Reviews: Climate Change*, 2(4), 547–569.

6 Teaching powerful geography through graphicacy, map work and visual literacy

Introduction

Living in a visual world, children make sense of their world through visual images and cues (Mackintosh, 2010). We live in a visual world. Images are a powerful tool of communication. Visual images are used to communicate ideas, to sell products, to question, to educate. They can be used for learning, recording and communicating ideas. While they can educate, inform and inspire, they can also manipulate our views by presenting a limited or biased perspective. As our exposure to images increases, their power becomes more pervasive. Due to rapid technological development and globalisation, visual literacy is part of the 21st-century learning skillset.

As so much information is communicated visually both within and beyond geography, it is important for children to learn how to be visually literate. Visual literacy is the ability to decode and interpret visual images. Children need to be able to read, interpret and critically evaluate images. Visually literate people can use pictures and images in creative ways. Mackintosh (2010: 122) argues 'the skill of looking at, understanding and interpreting pictures has to be taught through planned directed study'. However, children are given little guidance on how to read images (Hope, 2008; Danos, 2014). Practising visual thinking skills can be like detective work, putting the pieces together. Very often it is through the discussion around these activities that the geographical learning takes place.

Graphical intelligence refers to the ability to use graphic skills to communicate and interpret information, to solve problems and to add a visual dimension to data. This form of intelligence is of particular interest to geographers. Powerful primary geography has the potential to develop visual literacy, visual thinking skills and graphical intelligence.

This chapter sets out to:

- discuss visual literacy and graphicacy;
- highlight the value of field sketches;
- suggest techniques for working with images;
- outline the possibilities for creating geographical resources through taking good photographs and underline the importance of working with maps.

Visual literacy

Visual literacy refers to the ability to understand and produce visual messages. To be visually literate one needs to be able to create, use and evaluate visual resources. Resources for teaching visual literacy include images, graphs, drawings, paintings, sculpture, hand signs, street signs, international symbols, advertisements and much more. Visual literacy is particularly important in the context of geography. Learning to decode an image, comprehend it, understand its intended meaning and incorporate it into new knowledge is part of visual literacy.

Children today are global consumers of visual imagery and messages. In our contemporary world, heavily saturated with pictures and media, views of literacy must be extended, and redefined. Inundated with a steady stream of images from the internet, television, film and advertisements, children need to develop skills of discernment and critical analysis to deal with conflicting messages and misinformation. Social media applications including Facebook, Instagram, Twitter and WhatsApp present both opportunities and challenges for children. Children need to appreciate the power of visual images; they need to be able to read and understand these images and they need to appreciate whose story is being communicated and by whom. Conversely the opportunities for visual literacy provided by technology are substantial. Many children have access to devices with camera functionality. Children can create video messages and films and post them on the YouTube internet portal. They can send messages via smartphones, iPads and other devices. The web page Pinterest allows children to create thematic sets of images free of charge. Children can create their own visual stories, they can participate in conversations using visual images and they can learn the skills of critical thinking using visual images. Visual literacy is a critical 21st-century skill. As 21st-century learners and global citizens, children need to be able to critically read and decode visual images.

Graphicacy and graphical intelligence

Graphicacy, a form of visual literacy, is a key component of powerful geography. Graphicacy is the ability to understand and present information in a visual manner through media such as maps, graphs, diagrams and drawings. Understanding road signs, weather symbols, house plans and Google Maps all involve the skills of graphicacy. Graphicacy was the original form of communication between humans as documented in cave paintings which date to the Palaeolithic era 30,000 years ago. These drawings demonstrate evidence of map making, spatial planning and communication through symbols. The widespread use of graphics in our world today in newspapers, advertisements, government information pamphlets, political manifestos and on TV, demands a high level of competence in graphicacy.

The information conveyed through graphicacy can be directly representative of what we see (photographs and sketches) or more abstract (as in maps, plans and diagrams) or numerical (graphs and tables). Graphics offer three advantages: (1) they are concise – a photograph captures a scene in one image; (2) they are memorable – the London Tube map is a constant reference point for Londoners; and (3) they convey a range of spatial (maps) and non-spatial (graphs) information. Use of Web 2.0 facilities (including practices such as blogging, social media, podcasting, videocasting, digital storytelling and participation in online discussion) provides incredible potential for teaching graphicacy. Children from St Augustine's NS, Clontuskert use Web 2.0 tools extensively and examples of this are discussed in Chapters 1 and 5.

Spatial understanding involves appreciating where one object is located in relation to another. Learning to understand spatial relationships helps children to describe where things are located. Spatial concepts are explored when the following words are used: *above, below, before, after, high, low, in front of, behind, inside, outside, on top of* and *underneath*. Spatial intelligence includes the ability to incorporate mental imagery, spatial reasoning, image manipulation, graphic and artistic skills, and an active imagination. Sailors, pilots, sculptors, painters and architects all exhibit spatial intelligence. The ability to read spatial forms of information requires a range of graphicacy skills such as interpretation, decoding, map reading and the ability to understand symbols.

Spaces can be real, virtual or imagined. Children can explore real places in their local areas, virtual spaces with the assistance of ICT, or they can create their own imaginary spaces. Spatial thinking is an essential part of everyday life. Whether you are finding a car, retracing your steps, going on a journey, finding a room in a hotel, following a map, you are making spatial decisions. Spatial thinking includes: map reading; recognising symbols and signs; understanding landmarks; being able to use and navigate systems of roadways; and knowing that countries are divided into areas which may be geographic, economic, social and political. Spatial thinking also includes the use of frames of reference such as street numbers; spatial language such as *adjacent*, *left* and *right*; and orientation and direction concepts such as north and south. Spatial thinkers present data using graphic formats, they highlight data visually and they create 2-D maps based on 3-D environments.

Graphical intelligence refers to the ability to use graphic skills, more specifically 'the ability to integrate the use of eye, mind and hand to solve problems of various kinds and generate effective products aimed at creating, acquiring and communicating knowledge' (Cicalò, 2017: 6). Through the successful coordination of eye, mind and hand, that is, perception, thinking and representation, one exhibits graphical intelligence. This intelligence involves the ability to translate mental images into drawn images or sketches. Drawing is considered a cognitive process and a tool for thinking. Human beings have an innate ability to graphically represent concepts and images, but this remains an underdeveloped intelligence in schools and within teacher education. To promote this intelligence, children need to be exposed to a rich graphic environment, they need regular opportunities to construct and design maps and they need an opportunity to sketch/draw more often. Cicalò (2017) argues that drawing has to be considered a central rather than a peripheral cognitive ability. Drawing on research from the neurosciences, Cicalò argues that drawing activates new avenues of thinking and seeing. Powerful primary geography promotes graphical education.

Field sketching

Sketching is a powerful method of capturing and communicating spatial concepts. A common activity, sketching includes informal activities such as tracing a symbol in sand, drawing a simple map indicating directions and doodles on whiteboards to more formal activities such as drawings for engineering, architecture and interior design. Sketching involves 'the purposeful construction of visual representations, marks that can then be reinterpreted in thinking, learning, and communication' (Forbus and Ainsworth, 2017: 864). Field sketching (as discussed in Table 6.1) is important for recording geographical phenomena in local physical and human environments, such as local building developments, river landforms or seasonal changes. It allows children to annotate key features of the place being observed. Sketching by its nature involves time, effort, observation and attention to small detail. Attention to scale is important. Being an artist is not a requirement for field sketching. What counts is not the final product but the cognitive process which is triggered by the act of sketching (Cicalò, 2017). Sketching requires us to look at the entity in a concentrated way focusing on lines, shapes, relationships and details that generally escape the eye. Involving a high level of attention, sketching is essentially a fundamentally new way of seeing, as it changes the way the viewer looks. As we inattentively observe our daily surroundings 'the brain is satisfied with recognizing what it sees, without deepening the observation, interpretation and understanding of what the eye perceives' (Cicalò, 2017: 6).

Observational skills generally improve as children draw and annotate their sketches. Field sketches are an effective way of recording observations made during local trails (Figure 6.1). These trails or local walks may take place in the school grounds, in the area immediately surrounding the school or further afield. In addition to providing an image, field sketches can be used to record information when observing aspects of place. They can include a set of notes including name of author, location, time, weather conditions and reasons for compiling this particular sketch. Unlike a photograph, these specific details can be included for future reference. Field sketches also help the composer to record memories of the entity being observed. From an assessment perspective, sketches illustrate visual, spatial and conceptual understanding.

Table 6.1 Construction of a field sketch

Equipment: Viewfinder, clipboard, record sheet with a grid, pencil, compass

1. Study the place being observed and select features to be sketched. Choose a suitable viewing point. A viewing frame or a viewfinder (a cardboard rectangular frame) can assist children to select and frame their scene.
2. Decide on landscape or portrait layout. Prepare a grid. Advise the children to divide their scene into three parts. Any of the following options can be provided:

3-part grid divided into thirds	3-part grid divided into foreground middle ground and background	9-part grid

3. Using the compass, find the North (the direction towards which the compass needle points). With an arrow and the letter N, record the North on each sketch.
4. Using a pencil and blank sheet secured on a clipboard, children draw outline of their scene marking in prominent land features, the horizon and other significant lines in their viewpoint.
5. Continue to observe and mark in other features and smaller details. Shading and colouring can be used to identify key features such as shrubs and trees.
6. Label key features, include a title and annotate the drawing with date and name. Insert a key if using symbols, colour or shading.
7. A final frame is optional. Older children can scan their sketches and use them for PowerPoint presentations.

Figure 6.1 Children taking field sketches

Case study 6.1 Eyrecourt streetscape

Children (11–13 years) from St Brendan's NS, Eyrecourt, Co. Galway constructed a streetscape of their village. Working in pairs they stood in a line along the street with one building in front of them (Figure 6.2). In each pair, one child described features while the other child drew. A4 sheets of paper were distributed and the children were asked to divide each sheet into thirds. The first third was dedicated to the first storey of the building. The second was for any other storeys. The final third was designated for the roof. The children were asked to pencil draw an initial outline of the building, attending to minute detail. They then sketched in windows and doors. The children were asked to note the number of storeys and bays on each building. Having observed and noted the gutter line and ridge line on each roof, the children drew in window sills and noted colour and unusual brick layouts. Details on doors and windows along with signage were documented. Back in the classroom the children finished their sketches, completing final details and adding correct colours. The sheets were arranged on a larger sheet of paper in the same order as the buildings on the street. The final streetscape is a class creation proudly displayed in the school hall and the children were very pleased with their work (Figure 6.3). This activity was adapted from *Archaeology in the Classroom – It's About Time!* (Limerick Education Centre, 2005).

Working with images

A photograph is a powerful geographical resource. Photographs can help children understand geographical terminology, processes and locations in terms of human, physical and environmental geography. Images and photographs can be dynamic tools in the geography classroom. When working with images contemporary issues are widely available from local and national newspapers. Photopacks are available from non-governmental organisations such as Oxfam, Trócaire, Christian Aid and Concern. Local photopacks can be made from the children's photography work, as discussed later in this chapter. The following is a list of photography activities which teachers can use in school to promote geographical curiosity and engagement.

Figure 6.2 Working on the streetscape

Figure 6.3 A section from the streetscape of Eyrecourt

Discussing images and photographs

Question the children about the picture to elicit geographical information. Encourage the children to give reasons for their answers.

What time of the year is it? Which season?
Where is this place?
What work do people do in this area?
What can we learn about this place from the photograph?

What is the photograph telling us?
Describe what you see.
Where was the photograph taken? Describe locational clues.
Who are the people in the photograph?
What does the photograph tell us about their lives?
If this was in a newspaper what caption would you use?
How would you categorise this photograph?
What do you think happened before and after the photograph was taken?
Why did the photographer take the photograph? What was the photographer trying to convey? How might he or she have felt when taking the photograph?
What is your response?
Do you like or dislike the photograph?
What other images does this photograph remind you of?
What is the overall message of the photograph?

Describing the photograph

Describe this place under a number of headings such as: buildings, landscape, transport and people. To stimulate discussion and interest in a topic, select some photographs which refer to a geographical issue. Children work in pairs or small groups to draw up a list of questions about a photograph. These questions can be used as the basis for a class discussion.

Show the children a photograph. They are allowed to look for 30 seconds to 1 minute. After this time the photo is turned over and the children have to list all the elements they can remember.

Ask one child to describe a photo while another child draws the picture. Use a range of locational language to describe the location of different people and objects in the photograph e.g. *near, far away from, beside*. Compare the photograph and the picture and discuss the process. How easy is it to describe accurately? What could improve the instructions?

Questioning a photograph

Place a photograph on a large sheet of paper. Ask the children to write any questions they have about the image (Figure 6.4). Discuss the questions as a group and discuss similarities,

Figure 6.4 Questioning a photograph

patterns and categories, and suggestions for answering questions. Agree one enquiry question and use this as the basis for an investigation.

Responding to a photograph

Writing a caption: Individually or in pairs, children can write their own captions for some photographs. Alternatively, they could select the most appropriate caption from a list prepared by the class teacher.

Share responses to photos on sticky notes: Display a series of photographs on the classroom walls and ask children to write responses on sticky notes. Responses might begin with: 'I like this picture because … ', 'This picture makes me think about … '. Using superimposed 'thought bubbles' write what the person is thinking. Write speech bubbles for each person portrayed in photograph. Discuss the people in the photograph and try to imagine what each person in the photograph is thinking or feeling. Alternatively, a conversation inspired by the photograph can be recorded.

Beyond the frame

A photograph is a record of a frozen moment. To give the photograph some context children can place the photo on a large sheet of paper and draw around it to extend the story. They will need to discuss the possibilities before deciding on their picture extensions. Display the results and ask the children to share their decisions with the rest of the class. Alternatively, if you want to compare children's ideas with the reality shown in the photograph, you could reveal just a section of the original image and ask them to extend it. Their extended frames can then be compared to the original photograph. Using role play and improvisation children can imagine what happened before or after the photograph was taken. Some geographical conceptual information shared with the children will enhance all discussions and enquiries.

Collecting photographs

Children can collect photographs illustrating different geographical features (Table 6.2). Ideally these should include photographs of local features taken by the children themselves.

Table 6.2 Samples of physical and human features which can be illustrated in geographical photographs or images

Physical features		Human features	
Landscapes	Coastal, rural, mountain	Settlement	Cities, towns, suburban areas
Landforms	Headlands, islands, estuaries	Land use	Urban, agriculture, leisure
Vegetation types	Woodlands, bogs	Transport	Rail, roads, cycle paths
Climate	Temperate, temperature, evidence of change	Services	Shops, council, post office
Waterways	Rivers, streams	Population	Number, density, age and gender
Soil	Texture, colour, structure	Heritage	National parks, historical features and monuments

Sorting pictures

Ask the children in groups to sort their pictures into sets. You may provide some suggestions to begin with (those with water, homes or people). Then ask the children to work out categories of their own. Which place would they most like to visit and why? This can be a group response, which means children have to discuss and come to a majority decision.

Working with aerial photographs

Ask children to:

describe the physical and human characteristic of the region in the photos;
describe the relative location of familiar landmarks;
list the forms of transportation in the photo;
discuss how the land is used;
find residential, commercial, industrial, recreational, transportation and agricultural spaces;
source examples of how humans have modified the environment;
find evidence of weather and climate on the area;
show negative and positive effects of human–environment interaction;
select one physical or cultural landmark and discuss its current and possible future impact on the area;
predict what changes may occur in the area in the next five years with reasons for their suggestions.

Working as a photographer: How to take a good photograph

Photography has become accessible to schools and children as mobile devices such as tablets and iPads have embedded cameras. Interactive whiteboards allow children to share their images at no cost. The size and affordability of smaller cameras makes incorporating images into geography lessons easier than ever. It is worthwhile teaching children how to take a good photo as their images will enhance the school's geographical resources and the teacher's potential for incorporating the children's own geographies. Here are some guidelines:

- Using a cardboard rectangular viewfinder walk the local area and notice potential aspects of the local environment which could be photographed. Mentally compose some shots and consider the story you wish to tell.
- Stand with two feet firmly on the ground with camera held steady in both hands.
- Think about the composition of your image. In photography, composition refers to the arrangement of elements. It means paying attention to what will be photographed and how it will be positioned. Decide upon the key subject of your photo and fill the frame. Sometimes photographs are taken with the main subject residing in the bottom corner of the frame, which is not an effective use of layout. Using the rule of thirds helps to make a photograph feel more dynamic. Imagine your screen is divided into a grid of nine equal squares with two horizontal and two vertical lines (Figure 6.5). Some cameras display these composition lines. If you position your subject along these lines or at the points where the lines intersect the image will be more effective. Ask children to avoid placing their selected subject directly in the middle of their picture.

156 *Teaching powerful geography through graphicacy*

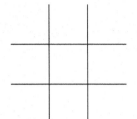

The Rule of Thirds

When looking through your camera imagine a tic tac toe grid. This grid divides your picture into thirds. Both the horizontal and vertical sections are split into three sections.

Learning how to focus is important for taking a sharp clear image. Most digital cameras have a central auto-focus facility indicated by a small square on the screen. To activate this facility simply half press the shutter button and hold. Once confirmation of the focus has been acquired (usually by a green icon displayed on the screen or view finder) press the shutter button to complete the process.

Some preliminary photography exercises for children

1. Ask children to photograph their pets, trees, leaves, flowers or some of their toys at home.
2. Encourage children to photograph an object from many different angles and points of view.
3. Ask children to look for repeated elements or patterns, e.g. steps, fence posts, blinds, to make them aware of repetition.
4. Encourage children to capture perspective by acquiring a line that travels through the photo with the help of a road, wall or fence.

Geographical photography: geo-photo challenge

Setting geo-photo challenges for children inspires excitement, enthusiasm and a reason to engage geographically with their local area. Outdoor observation skills are developed through continued practice. Children should be encouraged to use their senses when gathering

Figure 6.5 Rule of thirds: Picture of Giant's Causeway illustrates the aesthetic importance of good composition

Source: image supplied by Richard Greenwood

information through observation. When children see, hear, taste, smell and touch the world, they are gathering geographical information about how the world works. Learning how to sit/stand quietly in one place for ten minutes often enables more effective observation. Provide children with a special notebook and pencil for recording observations, and help them to write down notes as necessary. Observations can be captured and shared afterwards through photographs. Tables 6.3 and 6.4 provides examples of geo-photo challenges:

Table 6.3 Scavenger hunt for younger children

Take a photograph of ...	
Something tiny	Something red
Something large	Something with wheels
Something that moves	Something square
Something special	Something round
Something orange	Something old
Something green	Something new
Something smooth	Something in the sky
Something shiny	Something that begins with the letter 'S'

Table 6.4 Scavenger hunt for older children

Take a photograph of ...	
A specific colour in your local area, objects that are green, red, blue or yellow	An unusual sign
Shapes in your area; encourage the identification of 2-D and 3-D shapes	Local maps and evidence of map reading
	An example of people interacting with the physical environment
Evidence of geography in local signs	A food or drink in your local supermarket which comes from an unusual place
Weather-specific instances, e.g. snowflakes on leaves, raindrops on flowers, damage caused by a storm	
A car with a licence plate from a different county and/or country	

> *Exercise 6.1 Activity for teachers*
>
> 1 Using your camera or iPad take one photograph of something which inspires awe and wonder for you. Try to share an image which made you say 'wow'.
> 2 Share these photos with your staff and discuss the special elements and unique features of each photograph.
> 3 Discuss ideas for promoting and sharing children's awe and wonder.

Case study 6.2 Children working as photographers: A to Z of Eyrecourt

Children (8–10 years) from St Brendan's NS, Eyrecourt worked on an A to Z PowerPoint presentation based on their local area. Initially, I showed the children some A to Z picture books. Inspiration for this activity comes from alphabet picture books. There are a series of alphabet books published by Frances Lincoln in conjunction with Oxfam. These include

158 *Teaching powerful geography through graphicacy*

I is for India, C is for China, W is for World, A is for Africa and *J is for Jamaica*. Generally, there is a page for each letter of the alphabet, with text and photographs to illustrate aspects of life in that country. So, in the book *C is for China* (So, 2006):

A is for Abacus, a calculating frame with beads which slide on thin rods;
B is for Bicycle, the most common form of transport;
C is for China, one of the largest countries in the world;
D is for Dragon, a legendary animal which is a symbol of strength.

These books provide a stimulus for children to make their own picture alphabet books about different places (Russell, 2008). Children's work is showcased in Case Study 2.3 *Granapedia: The first ever encyclopaedia of Granard: The A to Z of Granard* (Figure 6.6).

Figure 6.6 Cover page from Granapedia: The first ever encyclopaedia of Granard: The A to Z of Granard

Teaching powerful geography through graphicacy 159

Figure 6.7 One child presents her A to Z project

Children from St Brendan's NS, Eyrecourt designed their own A to Z book about their locality, Eyrecourt. Children worked in pairs to produce these books. Each page/double spread typically includes images, some text and a map. Sources for research included the internet, reference books and personal experiences such as holidays. After an initial brainstorm the children decided on the focus for each letter. Difficult letters in the alphabet are included by adopting strategies such as *X Marks the Spot* to include a map of the place. This activity encourages children to think about their local area using the framework of the alphabet. When the children are outside looking for subjects to photograph it focuses their minds and helps them to see their place from a new perspective. The children presented their work to the senior classes in their school (Figure 6.7).

Feedback from student teachers

Student teachers are introduced to this concept by completing their own A to Z guide of the college. Their feedback is noted as follows:

> It made us think and it makes the local personal especially when we put ourselves in the picture.
> The last few letters are difficult and some of the final interpretations are abstract due to unique interpretations.
> It's a very creative way of getting to know our surroundings and it's a fun way of working in groups.
> Learning to navigate: which is the best route to get the task completed in the least amount of time.
> Everyone interprets the task differently as we discovered during the final presentations and it shows that you can use absolutely anything for geography.
> It's a simple way of working as explorers outside or inside if it is raining.
> Working outside is something you will always remember.

Case study 6.3 Picturing Places: Learning about Oranmore

Children (8–10 years) from Gaelscoil de hÍde engaged in a 'Picturing Places' project. Oranmore is a small coastal village located in County Galway in the West of Ireland. The aim of this project was to help children learn about their local area through fieldwork, photography and digital learning. Through this project, children conducted walks in Oranmore, took photographs of geographical features, prepared presentations and delivered these presentations in class (Figures 6.8 and 6.9).

The project entailed some short walks in Oranmore during the month of May. After conducting two sessions on the basics of good photography, the children were able to compose their own photographs and download their images. In a note to parents, an appeal was made for photography equipment:

> If you have the following equipment at home I would be grateful if you could give it to your child.
>
> - A digital camera or a tablet with a camera (note if the child is using a tablet could you ensure images are of a reasonably good quality and that you are able to help your child download images from tablet to memory stick).
> - A blank memory card for your camera.
> - A memory stick for storing photographs.
> - If you have a digital camera could you check and see if you have the equipment for transferring the images from your camera to your computer, e.g. a card reader may be useful.

NB Please ensure memory cards are cleared before they are brought to school.

Please do not buy this equipment as we do not want you to incur any expense. If you have this equipment at home please make sure it is labelled and that batteries are fully charged for your camera. If the children could have their digital equipment ready for Wednesday we can organise for children to share equipment if necessary. If you do not have this equipment, your child will be able to share with other children.

The project took place in four different stages as indicated in Tables 6.5, 6.6 and 6.7.

Table 6.5 Picturing Places project: learning about Oranmore: four phases

Phase 1: Preparations	Phase 2: Walking in Oranmore (part 1)	Phase 3: Walking in Oranmore (part 2)	Phase 4: Compilation of findings and presentation of work
Mé féin	Group 1: Shapes and patterns	Group 1: Transport in Oranmore	A to Z of Oranmore
Learning to use PowerPoint: short presentation 3–4 slides per child	Group 2: Numbers and maths trail	Group 2: Education in Oranmore	Seven Wonders of Oranmore
Myself	Group 3: Facilities for children	Group 3: Large shops in Oranmore: SuperValu and Oranmore town centre	Class map of Oranmore
My family	Group 4: Evidence of Irish language (Gaeilge) in Oranmore		
Where I live (meaning of place name)	Group 5: Links to other places	Group 4: Small shops in Oranmore	
	Group 6: Nature in Oranmore (our natural environment)	Group 5: Services in Oranmore	
Places where I like to visit and reasons for this	Each group will be asked to note the following if it arises: something surprising or unusual in Oranmore	Group 6: Visit to a factory in Oranmore (subject to permission)	

Phase 1: Preparations

The first part of the project involved children preparing and making a short presentation about themselves using 3–4 photographs. This part was designed to help the children learn how to take a good photograph and how to use the digital equipment. The children learnt how to use PowerPoint, and prepared a short presentation for their own class.

Phase 2: Walking in Oranmore (part 1)

The second phase of the project involved the children taking photographs in Oranmore in accordance with a particular theme, e.g. shapes and patterns in Oranmore (Table 6.6.). Each theme could be interpreted broadly by the children. Children were encouraged to

Table 6.6 Themes for each group (Phase 2)

Group 1: Shapes and patterns
Look out for interesting shapes and patterns in Oranmore. Look out for 2D and 3D shapes. Some ideas:

- Windows and doors
- Brick patterns/paving stones
- Roofing tiles
- Climbing frames
- Right angles
- Symmetry

Group 2: Numbers and maths trail

Number	Number – Steps and stairs (counting). Measuring and estimating distances. Plant growth. Repeating patterns.
Measures	Measures – Distances. Estimating length etc. Timing activities. Coordinates on a map (treasure hunt).
Data	Data – Graph making. Weather recording. Tallying – Recording birds that visit bird table.
Around the school/On the road	Around the school. On the road – Traffic and pedestrian surveys. Shapes on road signs. Distances on signposts. House numbers. Number plates.
Shops/Post office	Shops/Post office – Opening times. Price of a stamp. Posting overseas. Basic food items. Money handling. Change. Measuring a post box. Letter opening. Weight of parcels.

Group 3: Facilities for children
Look out for facilities for children e.g. playground and library. Ask the children to take pictures of things they like, don't like and which could be improved. Think about the children making recommendations for the county council about children's facilities in Oranmore.

Group 4: Evidence of Irish language (Gaeilge) in Oranmore
The children are currently undertaking their education through Gaeilge. They were asked to take pictures which highlight the use of Gaeilge in Oranmore e.g. bilingual signs such as those in Figure 6.7. The children were asked to take note of local shops which make an effort or which make no effort to communicate through Gaeilge. Children were also encouraged to listen to other European and international languages spoken in Oranmore.

(Continued)

Table 6.6 (Continued)

Group 5: Links to other places
Look out for any link to another place in Oranmore. Examples include signs, restaurants (Chinese\Italian) and maps. Are there businesses in Oranmore from other countries? Look out for links between Oranmore and other towns in Galway, Ireland and other places around the world. Have we people here from other countries? Find out some interesting stories. Take pictures of people if appropriate but remember to ask permission first.

Group 6: Nature in Oranmore (our natural environment)
Take pictures of nature/the natural environment in Oranmore. These can include pictures of the sea, green areas and local flora and fauna. Ask the children if they can name some of the trees and flowers. Take pictures of green areas and specific examples of nature e.g. an oak tree. If they are not able to identify specific trees and flowers, ask the children to take pictures and they can be identified later. Don't forget Galway Bay. Talk to the children about the significance of Oranmore being located so near the sea. During this walk we are investigating services and amenities for tourists and people who might be interested in settling in Oranmore. Each group is required to sketch one amenity/service and take photographs.

---

make notes when taking photographs, to describe the location and to outline details of the main feature in their image.

Parents accompanied children in groups of five. During their thematic exploration, each group was asked to note anything surprising, unexpected or unusual in Oranmore. If the children decided to take a photograph of a person, they had to ask for permission beforehand. Children were encouraged to ask questions about different physical and human features observed during their walk. Each group prepared a PowerPoint presentation based on their photographs and delivered it to a different class.

Phase 3: Walking in Oranmore (part 2)

The third phase of the project required the children to explore Oranmore through a particular geographical lens such as transport in Oranmore (Table 6.7).

Table 6.7 Themes for each group (Phase 3)

Groups 1 and 2: Transport in Oranmore
Examine how people travel around Oranmore and to other areas. Visit the bus stops and look at timetables. Visit the train station and count the number of cars in the car park. Examine the timetables. If possible talk to a person who has used the train. Is the train schedule sufficient? What could be done to improve the train station? Where can one travel to in Ireland by train?

Examine the issue of traffic in Oranmore. Where are the problem spots? Is there a cycle lane? Is there any place to leave bicycles in Oranmore? What could be done to improve traffic in Oranmore?

Groups 3 and 4: Education in Oranmore
During this walk we are investigating services and amenities for tourists and people who might be interested in settling in Oranmore. Each group is required to sketch one amenity/ service and take photographs. Take photographs of the three primary schools and walk to the secondary school. Find evidence of other types of education in Oranmore, e.g. crèche, preschool, library and adult education. Try and establish any significant differences between the schools. If possible you might meet someone who could show you around.

Groups 5 and 6: Large shops in Oranmore: SuperValu and Oranmore town centre
During this walk we are investigating services and amenities for tourists and people who might be interested in settling in Oranmore. Each group is required to sketch one amenity/service and take photographs. Go to the Oranmore town centre and establish what facilities are there, e.g. post office. Talk to a member of staff in SuperValu.

Phase 4: Compilation of findings and presentation of work

In the final stages of this project the children used their photographs, sketches and images to compile an A to Z of Oranmore. They voted on the Seven Wonders of their area and finally they constructed 3-D and 2-D maps of Oranmore.

Figure 6.8 Evidence of the Irish language in Oranmore

Figure 6.9 Picturing Places project: Learning about Oranmore

> *Exercise 6.2 Activity for teachers*
>
> As a CPD activity invite all teachers to work outside to compile a number of field sketches of the local area. What were the common learning points from this activity?
>
> Engage in some of the geo-photo challenges outlined in this chapter. Share your work with the children and invite their responses.
>
> Begin a collection of inspiring images from local and national newspapers. Display a *Geographical Image of the Week* on a school noticeboard. Invite comments from teachers and children.

Working with maps

I once visited a school in Salzburg, Austria, and was introduced to a class of 10-year-old children. The children immediately opened their atlases to locate my places of birth and work in Ireland. I was struck with their familiarity with the atlas and their regular use of it as a reference source. One of the most important aspects of teaching map skills is access to a broad range of maps in primary schools. At a very basic level classrooms should have a globe, a map of the world, a map of their country, local maps and an atlas for each child.

Maps are becoming increasingly popular in contemporary society. The use of Google Maps for locating places and planning routes is expanding rapidly due to widespread availability on mobile devices. Cartography and geo-information systems are becoming increasingly important in economic, social and political planning and innovations. Yet Wiegand claims that 'mapping in schools has not kept pace with developments in cartography most of which have been computer based' or in other words 'map related pedagogy is poorly developed' (2006: 2). Considering the centrality of maps and map skills to geography, this is a disappointing claim.

A map is a two-dimensional representation of a three-dimensional space. Maps use symbols to represent different kinds of data. Maps are perhaps one of the oldest forms of non-verbal communication in the world. They illustrate spatial relationships in terms of categories such as land use, population, economic activity, physical features and buildings. It is important to remember that maps convey only partial information. Once children begin to make their own maps they become aware of this partial nature in terms of what they decide to include and exclude.

Sobel (1998) considers map making to be a crucial skill in primary schools. Map reading skills are part of visual literacy. Maps present particular spatial data and symbols which children have to learn to read. They provide a valuable bridge between real and abstract worlds. They help to develop a sense of place. Children's maps and drawings include representations of places, events and things that are emotionally important. Working with 3-D materials is a very accessible form of map making for children. Hart (2013) noted that children's maps made with 3-D materials were more accurate than their 2-D drawings. Google Maps now helps to bridge the gap between the children's real 3-D world and a 2-D representation by combining satellite and street view imagery.

Case study 6.4 Scoil Íde climate change: Problem-solving through mapping the local wetlands: A quest to measure, map and log data on Bishop's Field wetland

As part of their geo-literacy programme, teachers from Scoil Íde, Corbally prioritise map work and map skills. An essential component of geo-literacy is the understanding of maps, which is often coined 'map literacy.' Sixth class children (11–13 years), along with teacher Neil Foley, conducted a very interesting project in mapping the seasonal wetlands in the Bishop's Field beside their school. The Bishop's Field is an open, part-wooded green area in the suburban area of Corbally, situated between Scoil Íde primary school and St Munchin's College secondary school. It slopes gently from west to east where it flattens out to a slight depression. Part of the Bishop's Field is used by the primary school for rugby, games and athletics. The wooded area is visited frequently by classes on nature field trips. For the past few years the children have noticed that a large low-lying area of the Bishop's Field has flooded each year during the wettest months, usually December to March.

Enquiry question: Why does flooding occur in Bishop's Field and does it affect flora and fauna?

The children examined the formation of this wetland and addressed their enquiry question by:

- mapping the relief of the land;
- measuring and logging the depth of the water (weekly);
- measuring the area of the wetland by grid system and aerial photos (monthly);
- correlating weather variables (rainfall, wind strength, temperature) with changes in the wetland;
- looking at factors that have contributed to the formation of the wetland in recent years – asking whether rainfall has increased during winter months and whether there are other factors in the local human environment which have contributed to the appearance of this wetland;
- recording changes in flora and fauna due to the wetland.

A grid system was created and stakes were inserted to track the seasonal variations in plant and animal life as well as rising and falling water levels. With the help of a parent, the children had access to photos and video taken by a drone. The children kept a blog whereby they shared their learning experiences. The geographical and cross-curricular learning from this project were impressive. Children had an opportunity to collect and analyse real-life data. They then went on to share their findings. Information was shared widely on Twitter and other social media platforms. The project featured in local and national newspapers and the children presented their work as part of the National Primary Science Fair (Figure 6.10).

Case study 6.5 Journey sticks

Aboriginals in Australia and Native Americans are said to have used journey sticks to tell stories about their travels. In many countries people have developed the idea of creating

Figure 6.10 Mapping the local wetlands

Source: image supplied by Neil Foley, Scoil Íde, Corbally, Limerick

a journey stick to help them tell the story of their journey to others. It involves tying to a stick objects and colours that represent different experiences, feelings or parts of the journey.

When Australian Aboriginals went on a journey they collected things and tied them to a stick in chronological order. After a long time they finally returned to their people. Referring to the objects attached to their sticks, they were able to remember their journey and recount their stories. This formed a verbal map which described the journey to someone who wasn't there.

This personal way of recording journeys has been adapted as a geography activity for children (Whittle, 2006). Journey sticks are created when children walk in an area. They collect items of interest and attach them to the stick (with wool or sticky-tape) in chronological order. As the items and the sticks are personal to every child this creates a 'personal geographical experience' (Whittle, 2006: 11).

Preparation: Provide sticks, sticky-tape/wool/string, cards for the children. This can be done as a paired or group activity. Use digital cameras to record the journey and observe the local environment that the children are working in. As well as giving children the opportunity to discuss objects that they may not be able to pick up, e.g. mushrooms, this also promotes effective cross-curricular links with ICT.

As children walk around the woodland/outdoor area, ask them to collect objects that they find in the area which will act as a visual prompt of their journey in the wood. This could include: feathers, leaves, litter and flowers. Children can also make a sketch on a card to remind them of a particular point of their journey. During the journey (or at the end of the journey for younger children) ask the children to attach the objects and/or drawings to the stick using the string/wool/tape.

After the walk: At the end of the walk or in the classroom, ask children to use the objects that appear on their stick to talk about their journey. Children can make a map based on their journey sticks (Figure 6.11).

Student teachers are generally very positive and enthusiastic about making journey sticks as indicated in the comments in Table 6.8.

Nevertheless they are slightly nervous about children making journey sticks in their classes for the reasons indicated in Table 6.9.

Table 6.8 Reflections by student teachers

- There's more to geography than reading a book.
- You notice your surroundings more.
- This activity is a great way to explore and get to know your local area.
- It's good for all kinds of learners especially those who learn with their hands.
- It makes us talk; we had a chance to use geographical words such as key, legend; it's great for working in groups and learning how to compromise.
- We were constructors of our own learning.
- It's a break from normal class teaching and it's practical.
- It gives great purpose to a walk and it's a wonderful opportunity to work independently.
- It promotes responsibility and provides a chance to think about what is harmful to the environment.
- Super introduction to map work.
- It's both active and interactive and it's great for communication skills.
- I felt like a child again on a treasure hunt.
- I didn't realise there was so much nature here.
- Personal memorable engagement in fresh air.

Table 6.9 Concerns raised by student teachers

- Afraid the children would run around uncontrollably which would cause discipline issues.
- I would be nervous about conducting this activity in an area unfamiliar to me.
- A significant number of adults would be needed to conduct this exercise safely therefore for practical reasons I cannot see myself conducting this activity.
- There are significant health and safety issues, for example, the children could pick up anything.
- This exercise would take up far too much time.
- The children could be easily distracted.
- Difficult to monitor the children.
- Because of the unpredictability of the Irish weather, it would be very difficult to plan for this activity.

168 *Teaching powerful geography through graphicacy*

In order to appease concerns the following advice is given:

1 In order to build up your confidence, conduct this exercise with a small group of children initially and assess its value.
2 When bringing more than one group, bring as many adults as you need. Parents are often happy to oblige. Older children can be very helpful if you are trying this exercise with younger children.
3 Give the children very clear instructions about what they can and cannot pick up. Alert them to the danger of broken glass and items with rough edges.
4 Some students have stated there is nothing for children to pick up. In this instance children can sketch small drawings on pre-prepared pieces of paper and tie these to their journey sticks. For younger children small symbols can be left in tins to mark significant points on their journey. These can be attached to the journey sticks.
5 Finally, to reinforce the positive benefits of this activity, read the letter from an enthusiastic student published on pages 73–74 in Chapter 3 of this book.

Case study 6.6 Mapping Limerick project

The Limerick Smarter Travel School Project is working to support schools in the delivery of tailored school travel plans. This aims to promote more sustainable travel choices for the school journey. A school travel plan is a document produced by the whole school community and any other interested parties. It looks at how children, staff and visitors travel

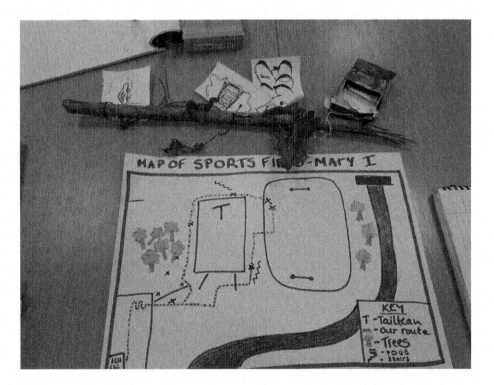

Figure 6.11 Journey sticks

to and from school and it suggests measures to encourage the school community to walk, cycle or avail themselves of public transport networks.

The purpose of a travel map is to enable children to become familiar with their local area and identify travel options available to them. A travel map should include:

1   Walking routes (key routes leading from the school);
2   Cycling routes and cycle parking (if in place);
3   Places of interest to children (local amenities, e.g. parks, shops, pool, library);
4   Landmarks in the area (if any);
5   Public transport networks (bus stops and routes);
6   Sustainable travel infrastructure (pedestrian crossings etc.);
7   Park-and-stride locations. These are places where parents could park and walk the remaining distance to school.

Conducting a walkability audit

A walkability audit is a methodical examination of how pedestrian-friendly a route is. The word 'walkability' means the 'ability to walk'. One aim of the Green Schools travel theme is to promote sustainable travel to school across Ireland. Raising awareness of road safety and sustainable travel also plays a large role in the Green Schools theme. St Patrick's Girls' National School is situated on the Dublin Road in Limerick City (R445). It currently has an enrolment of 225 children. The school is located on an extremely busy road. Approximately 1,000 vehicles an hour pass the school. A walkability audit was conducted by walking all the routes near school.

Objectives:

To look at the barriers to promoting walking to school and propose possible solutions if any.
To raise awareness of road safety.
To see if there are any park-and-stride locations by the school.

The children used a number of tools and resources on the audit to record their findings:

Recording sheets;
Measuring tape;
High-visibility vests;
Camera;
Pencils/clipboards/covers;
Analysis sheets/survey sheets.

Summary of findings (all of the findings were accompanied with photographic evidence)

Positive points

1   For safety, we have railings around our school.
2   We discussed how to cross the road safely.
3   We looked for safe places where parents and children could meet each morning and walk to school together in a walking bus.

Negative points

1. Double parking and traffic congestion are a real problem.
2. The signage near our school is not adequate.
3. One particular junction is wide and busy and difficult to cross safely.
4. We timed the traffic lights. The first lights took 62 seconds to change. The second set of lights took 4.5 minutes before we could cross. We did not have enough time to cross before the lights changed again.
5. We noted damaged footpaths and will report them to the City Council.
6. We recorded any hazards and litter we found on our way.
7. Our school does not have any parking facilities. Cars are regularly parked along the footpaths adjacent to our school. Due to the absence of parking facilities for our school, illegal parking on the school route continues to be an issue.
8. Due to the overhanging branches the path is very narrow.
9. When the footpaths ended abruptly, we had to find safe places to cross the road.
10. We noticed much discarded litter, especially used coffee cups and dog poo.

Recommendations

1. Flashing lights and more visible signs would be better.
2. Some road junctions are very dangerous.
3. Overhanging branches need to be cut back so we can use footpaths safely.
4. Road signage needs to be improved as 500 children and their families cross this road every morning and afternoon.
5. We noticed that the footpaths and pedestrian crossings need to be repaired and the road markings redrawn.
6. Could the speed limit outside the school be restricted to 30km/hour?
7. The committee suggested that road signs with flashing lights would be more prominent and would warn motorists that they are approaching a busy junction beside two large schools.
8. Children who walk, scoot or cycle to school from Dromroe, Drominbeg, Angler's Walk and Rhebog Meadows must travel under a railway bridge which has no pedestrian access.
9. The corner outside our school is a litter blackspot. Perhaps some rubbish bins and colourful floral planters would help improve this area.

Following the walkability audit the following improvements were made:

1. Some speed ramps were constructed.
2. Green Schools signs with *Children Crossing* were erected.
3. Visibility at one dangerous junction was much improved.
4. The wait time for the traffic lights at the pedestrian crossing was halved.

Sustainable travel is a wonderful theme for schools, a theme which has endless possibilities to link to all aspects of the curriculum (Figure 6.12). One of the fundamental aspects of this theme is that it allows children to explore their local environment and implement actions that can have a real and positive impact on the entire school community. As part of Green Schools teachers are encouraged to explore travel as a curricular theme and use

Figure 6.12 Sample of children's work from St John the Apostle NS, Knocknacarra exploring sustainable travel

this work to improve routes to school with a view to encouraging more children to walk, cycle or avail themselves of public transport (Figure 6.13). The identification of problems associated with traffic congestion and illegal parking is a good starting point (Figure 6.14).

Some mapping ideas that centre on the idea of transport

1 Local area map illustrating the findings of a walkability audit (positive/negative features with regard to walking/cycling).
2 Road Safety Map: Local area map highlighting safety concerns and places where children would have to stop, look and listen.
3 What I see on my journey to school map (placing pictures/illustrations of what children see on the way to school).
4 Where I come from: Invite children to mark where they live (geographical distribution of children).
5 How I travel to school: Map how children travel with symbols relating to the various modes.
6 Map local walking routes.
7 Townland maps in a rural environment/Estate maps in the urban environment.
8 Distance/time maps – create a journey time map, i.e. identify set distances from the school, e.g. 0.5km, 1km, 1.5km and draw circles (of those radii) centring on the school. Colour-code the circles according to distance (5 mins, 10 mins, etc.).
9 School Travel Map: A map detailing travel networks in your school's local area, i.e. public transport options, park-and-stride.

Figure 6.13 Children from Gael Scoil de hÍde, Oranmore focusing on sustainable travel

10 Are the routes to school safe? Conduct a walkability audit with your children to find out. Discuss findings and associated issues. Consider writing a letter to the local authorities.
11 Where do children travel from and how?
12 What would be the quickest way to travel from your school to a local landmark (use Google Maps)?

Children's literature and map work

As well as the real world, children's imagined and virtual spaces provide valuable starting points for developing children's graphicacy skills. Maps in children's literature represent both real and imagined places. They are useful devices for helping the child

Figure 6.14 The walkability study illustrated the problems of traffic congestion and illegal parking

to put the story into perspective. Even when a map is not provided the words and illustrations help create a mental image which children can express in their own maps based on the story. Maps from children's literature can help children understand map conventions such as symbol, key and orientation. They can also be used as a stimulus for map-making skills. Examples include maps of the semi-fictitious Scottish island of Struay in the *Katie Morag* books and the map of Treasure Island. I have argued that picture books in particular offer children an opportunity to develop a sense of place and to express their feelings about place (Dolan, 2016). Maps can also be used to record children's knowledge and understanding of a particular story. Children can draw a map after reading stories such as *Little Red Riding Hood*, *Hansel and Gretel* or *Goldilocks and the Three Bears*. Older children can draw their own map of Hogwarts based on *Harry Potter and the Philosopher's Stone* or a map of the Shire (Middle-earth) in J.R.R. Tolkien's *The Hobbit*.

Hennessy and Joyce's book *The Once Upon A Time Map Book* is a collection of maps from six fictional fantasy worlds: the Cloud Kingdom from Jack and the Beanstalk, Aladdin's Kingdom, Snow White's Enchanted Forest, Wonderland, Neverland and the Land of Oz. Each double spread represents the settings from a story in the form of a map. On each map, a compass, quadrants with letters and numbers, and a key with local routes help children navigate their way. Children are invited to follow the route taken in each story as well as search for hidden objects hidden throughout. There are also little visual jokes hidden in each of the maps. For instance, in Wonderland two porcupines can be seen escaping the Queen of Hearts' Gardens, while two field mice pour a liquid down the throat of a passed-out lizard. Looking for visual clues helps children to look closely at an image or a map. The sister book *The Scary Places Map Book: Seven Terrifying Tours* (Hennessy and Madrid, 2012) invites children to explore scary places such as Transylvania and Nightmare House. As the children follow the maps and special instructions for locating special objects they have to avoid pitfalls.

The end-paper to *Winnie the Pooh* by A.A. Milne, first published by Methuen in 1926, is a map of the 100 'aker' wood (Hundred Acre Wood). Using the device of a map or bird's-eye view, it shows each of the characters in their own home or locality in the Hundred Acre Wood, where the story is set. When E.H. Shepard first drew the map of the 100 'aker' wood he based the location on Ashdown Forest, near the home of the Milnes in East Sussex. Over the years there has been much speculation about the location of the story and how many of the landmarks were inspired by real places. According to Christopher Robin in his own autobiography *The Enchanted Places*, Ashdown Forest is 100 Acre Wood. The best-known landmark is Poohsticks Bridge, which on the local map is called Posingford Bridge. Playing poohsticks on any bridge is great fun.

The *Mapmaker Chronicles* is an exciting adventure series about a race to map the world within one year and a boy who discovered more than he ever imagined. Written by A.L. Tait, this fantasy adventure series is for primary children aged 9–13 years. The four books are *Race to the End of the World*; *Prisoner of the Black Hawk*; *Breath of the Dragon* and *Beyond the Edge of the Map*.

The king is determined to discover what lies beyond the known world, and has promised a handsome prize to the ship's captain who can bring him a map. To do that, they'll need mapmakers – and 14-year-old Quinn is chosen because of his photographic memory and his ability to read and write. The journey is a race to discover what lands are beyond the narrow maps that his people have so far produced. As they proceed across the ocean on the ship, the *Libertas*, the crew face hostile inhabitants, monsters of the ocean, and the treachery of competitors, but Quinn's character and confidence grow as the months pass. This is a work of fantasy, but it is also a work of historical fiction since it has strong parallels with real events in history. The exploration of other continents and the discovery of unknown tribes and lands by European explorers between the 16th and 19th centuries are clearly paralleled in these stories. These books provide a useful context for discussing children's knowledge of maps, their importance, different types of maps, the role of a cartographer, tools used for navigation today and the ancient belief that the world was flat and that one would fall off the edge if one travelled too far.

Pat Hutchins' book *Rosie's Walk* features the hen enjoying a leisurely walk around the farm, completely oblivious to her stalker, the fox, who unsuccessfully navigates the farmyard obstacles. In the story Rosie walks across the yard, around the pond, over the haystack, past the mill, through the fence, under the beehives and arrives back in time for dinner. In Scoil Íde, Corbally, *Rosie's Walk* is read to the children a few times before the book is used as a basis for mapping activities.

The children (4–6 years) discuss relative locations of landmarks on the map, using positional vocabulary words such as *next to*, *up*, *down*, *right*, *left*, *across*, *between*, *toward*, *away*, *near* and *far*. The teacher focuses on the words the author uses to show the hen's movement from place to place on the farm: *across*, *around*, *over*, *past*, *through* and *under*. Children use a map and cut-outs of the story's characters to follow their actions through space and time. The children act out these movements with the cut-out of the hen. Using a fox cut-out the children describe and act out the fox's movements: his collision with a rake, his fall into the pond, his tumble into a haystack, followed by a burial under flour, and a nasty encounter with beehives. They also demonstrate the positional vocabulary words using real obstacles in the classroom. The story is re-told in the children's own words, using the map and positional vocabulary. Finally, the children make their own maps based on the story (Figure 6.15).

Figure 6.15 A map of Rosie's walk drawn by a 5-year-old child

What is so powerful about graphicacy, map work and literacy

The reading and interpretation of images and maps is an important element of geographical education. The level of visual literacy combined with the prior knowledge of the child determine how well an image or map can be created, read and understood.

Since the launch of Google Maps, Google Street View and various mobile applications, remote sensing images have become popular and easily accessible. Visual literacy for working with these resources cannot be assumed; it has to be explicitly taught. Making 3-D models of local areas (Figure 6.16) helps children to conceptualise spatial information and images available from Google Maps.

Children taking their own photographs for use in classrooms is an excellent example of children constructing geographical knowledge. Teacher support and guidance is essential for maximising the learning potential from the children's images. Drawing sketch maps is an active experience through which children develop their own understanding of place. Any map-making exercise promotes children working as cartographers and enhances geographical knowledge creation. Using the framework illustrated in Chapter 1, Table 6.10 demonstrates how teaching graphicacy, map work and literacy generates powerful geographical knowledge.

Figure 6.16 Mapping places through 3-D art

Table 6.10 Teaching graphicacy, map work and visual literacy as powerful geographical knowledge

Maude's five types of powerful knowledge	21st-century competencies	Teaching graphicacy, map work and visual literacy
Knowledge that provides children with new ways of looking at the world	Knowledge building	Constructing and reading maps, photographs taken by children used as classroom text, creating A to Z books of local, national or international places or issues
		Developing competencies in spatial thinking; includes visualising, interpreting, and reasoning using location, place, distance, direction, relationships, movement, and changes in space
Knowledge that provides children with powerful ways to analyse, explain and understand the world	Critical thinking	Asking geographical questions, appreciating that images and maps include and exclude certain elements depending on the creator, using spatial concepts, skills and spatial language to articulate thoughts and ideas
Knowledge that gives children some power over their own knowledge	Real-world problem-solving	Using map reading skills to solve local issues, e.g. propose alternative routes to alleviate traffic congestion, choose alternative sites for a new playground and conduct walkability surveys
Knowledge that enables children to follow and participate in debates on significant local, national and global issues	Collaboration and teamwork	Working in groups to create and discuss images and maps
		Creating a class book of images and/or maps as an example of collaborative learning within and beyond the class
Knowledge of the world	Global competency	Sharing children's images and maps with a broader audience through social media, local newspaper or national competitions
		Using images developed by children to illustrate schools' geography policy and approach
		Learning about where different cities and countries are located helps children gain a larger world view

Conclusion

Recent research in the field of cognitive sciences suggests the existence of important relationships between graphic representation and cognitive development (Cicalò, 2017). Geography is a visual discipline. Working with a variety of visuals is an integral part of primary geography. The need to be able to critically read an image has never been more important, considering the widespread appeal and dissemination of visual images through conventional and social media. Technology and multimedia (text, sound and image) change the way in which we are communicating in a globalised world. Images such as photographs, illustrations, graphs and maps are often used together with text in order to enhance meaning.

Children love to work with maps. Schools are strongly encouraged to invest in a robust collection of atlases, globes, OSI maps and a map of the world for every classroom. Map skills contribute to children's ability to read and understand images. A strong foundation in spatial thinking and associated vocabulary will serve as a framework throughout their lives, creating the geographical prompts necessary to truly process, understand and navigate the world.

Exercise 6.3 Reflection for teachers

What approach is taken in your school for teaching visual literacy, graphicacy and map skills?

Assess the collection of maps in your school. Does it need to be updated?

Have all classes access to basic maps including a map of the world, a map of your country and of the local area?

Further resources

Children's books

There are several fiction and non-fiction books for children which provide insights into maps and mapping, such as:

Banyai, I. (1998) *Zoom*, Puffin Books.
Banyai, I. (1998) *Re-zoom*, Puffin Books.
Bolt, Simmons L. (2018) *How to Read a Map (Understanding the Basics)*, Child's World.
Boswell, K. (2013) *Maps, Maps, Maps*, Capstone Press.
Boothroyd, J. (2013) *Map My Room*, Lerner Classroom.
Boothroyd, J. (2013) *Map My Neighbourhood*, Lerner Classroom.
Boothroyd, J. (2013) *Map My Country*, Lerner Classroom.
Boothroyd, J. (2013) *Map My State*, Lerner Classroom.
Cherry, G. and Haake, M. (illus) (2015) *City Atlas: Travel the World with 30 City Maps*, Frances Lincoln Children's Books.
Chesanow, N. and Iosa, A. (illus) (1995) *Where Do I Live?* Barron's Educational Series Inc. US.
Dillemuth, J. and Wood. L. (illus) (2015) *Lucy in the City: A Story about Developing Spatial Thinking Skills*, Magination Press.
Dillemuth, J. and Wood. L. (illus) (2017) *Mapping My Day*, Magination Press.
Dodson Wade, M. (2003) *Types of Maps (Rookie Read About Geography)*, Children's Press.
Elliot, H. (2013) *Henry's Map*, Philomel Books.
Fanelli, S. (1995) *My Map Book*, London: ABC.
Greve, M. (2009) *North, South, East and West*, Little World Geography.
Hartman, G. and Stevenson, H. (illus) (1993) *As the Crow Flies: A First Books of Maps (Rise and Shine)*, Simon & Schuster.
Hawkins, C. and Hawkins, J. (2006) *The Pirate Treasure Map: A Fairytale Adventure*, Walker Books.
Hennessy, B.G. and Joyce, P. (illus) (2010) *The Once Upon a Time Map Book*, Candlewick Press.
Hennessy, B.G. and Madrid, E. (illus) (2012) *The Scary Places Map Book: Seven Terrifying Tours*, Candlewick Press.
Hutchins, P. (1971) *Rosie's Walk*, Aladdin.
Jenkins, S. (2003) *Looking Down*, Houghton Mifflin.
Knowlton, J. (1986) *Maps and Globes*, Harper Collins.
Leedy, L. (2003) *Mapping Penny's World*, Square Fish.

Letherland, L. (2014) *Atlas of Adventures: A Collection of Natural Wonders, Exciting Experiences and Fun Festivities From the Four Corners of the World*, Wide Eyed Editions.
Miles, J. (2016) *Ultimate Mapping Guide for Kids*, Firefly Books.
Mizielinska, A. and Mizielinska, A. (illus) (2013) *Maps*, Big Picture Press.
Mizielinska, A. and Mizielinska, A. (illus) (2014) *Maps Activity Book*, Big Picture Press.
Mizielinska, A. and Mizielinska, A. (illus) (2016) *Maps Poster Book*, Big Picture Press.
Mizielinska, A. and Mizielinska, A. (illus) (2016) *Maps Special Edition*, Big Picture Press.
Olien, R. (2012) *Map Keys*, Scholastic.
Olien, R. (2012) *Looking at Maps and Globes*, Scholastic.
Priceman, M. (1996) *How to Make an Apple Pie and See the World*, Random House Inc.
Rabe, T. and Ruiz, A. (illus) (2002) *There's a Map on My Lap! All about Maps*, Cat in the Hat Learning Library, Random House Books.
Ritchie, S. (2009) *Follow that Map! A First Book of Mapping Skills*, Kids Can Press.
Shulevitz, U. (2008) *How I Learned Geography*, Farrar, Straus and Giroux.
Singer, M. and Lessac, F. (illus) (1993) *Nine O'clock Lullaby*, HarperCollins.
Smith, S. (2017) *Map Maze Book*, Usborne Publishing Ltd.
Sweeney, J. and Cole, A. (illus) (1998) *Me on the Map*, Random House USA Inc.
Treays, R. (1999) *My Street Young Geography Series*, Usborne Publishing Ltd.
Wheatley, N., and Rawlins, D. (illus) (2009) *My Place*, Walker Books.
Wise Browne, M. and Pizzoli, G. (2017) *North, South, East, West*, Harper Collins.

Web resources

Online there are several resources for teaching graphicacy, maps and map skills:

Barbara Petchenik Children's World Map Drawing Competition: biannual mapping competition: http://icaci.org/petchenik/
Digimap for Schools: www.digimapforschools.edina.ac.uk
Gapminder: www.gapminder.org/
Google Maps: www.google.co.uk/maps
Geocaching: www.geocaching.com/play
Open Street Map: www.openstreetmap.org
Ordinance Survey Ireland (Ireland's National Mapping Agency): www.osi.ie/
Ordinance Survey UK (Britain's Mapping Agency): www.ordnancesurvey.co.uk/
Map Zone: www.ordnancesurvey.co.uk/mapzone/map-skills/measuring-distance
Meaningful maps: A social portrait through the eyes of young people: http://meaningfulmaps.org/
National Geographic Online Atlas: http://ngm.nationalgeographic.com/map/atlas
Oxfam Mapping our World: www.oxfam.org.uk/coolplanet/mappingourworld.
Peters Projection of the World: www.oxfordcartographers.com/our-maps/peters-projection-map/
Quickmaps: www.quickmaps.com
Scoilnet Map: https://maps.scoilnet.ie
OS Mapzone: www.mapzone.ordnancesurvey.co.uk/mapzone
Ubuntu: Using photographs in development education: www.ubuntu.ie/our-work/ipps/using-photos/using-photos.pdf

Further reading

Catling, S. (2002) *Placing places*. Sheffield: Geographical Association.
Catling, S. (2010) *Mapstart 1, Mapstart 2*. London: Collins.
Danos, X. (2014) *Graphicacy and culture: Refocusing on visual learning*. Shepshed, Leicestershire: Loughborough Design Press Limited.
Hope, G. (2008) *Thinking and learning through drawing in primary classrooms*. London: Sage.

Mackintosh, M. (2010) Using photographs, sketches and diagrams. In Scoffham, S., ed., *Primary geography handbook*. Sheffield: Geographical Association, 120–133.

Mackintosh, M. (2017) Representing places in maps and art. In Scoffham, S., ed., *Teaching geography creatively* (2nd ed.), London: Routledge, 76–87.

Miller, L. ed., (2017) *Literacy wonderlands: A journey through the greatest fictional worlds ever created*. London: Modern Books.

Roan, M. and Taylor, M. (2012) *The little book of maps and plans*. London: A & C Black.

Sobel, D. (1998) *Mapmaking with children: Sense of place education for the elementary years*. Portsmouth: Heinemann.

Wiegand, P. (2006) *Learning and teaching with maps*. London: Routledge.

Articles published in the Geographical Association's journal *Primary Geography*

Barlow, A., Potts, R. and Whittle, J. (2010) Messy maps and messy spaces. *Primary Geography*, 73, 14–15.

Catling, S. and Baker, P. (2011) Wish you were here? Exploring postcard maps. *Primary Geography*, 75, 12–13.

Channer, K., Lynch, H., Mayor, N., Pett, F. and Rotchell, E. (2017) How do maps and globes represent our world? *Primary Geography*, 94, 16–17.

Chave, C. (2011) Mapping the British Isles with heart and head. *Primary Geography*, 75, 14–15.

Clarke, J. and Edwards, M. (2010) Map sandwiches: Creating digital maps from layers. *Primary Geography*, 72, 24–25.

Eyre, G. (2010) Photos in global learning. *Primary Geography*, 71, 20–21.

Halocha, J. (2008) Progressing through pictures. *Primary Geography*, 65, 24–26.

Mackintosh, M. (2011) Graphicacy for life. *Primary Geography*, 75, 6–7.

Martin, F. (2005) Photographs don't speak. *Primary Geography*, 56, 7–11.

McGregor, S. (2012) Wishing for gold around the world. *Primary Geography*, 77, 28–29.

Mycock, D., Norman, W. and Pickering, S. (2012) Blind futsal: The beautiful, geographical game. *Primary Geography*, 77, 20–21.

North, W. (2006) Visual literacy, ICT and geography. *Primary Geography*, 60, 32–34.

North, W. (2008) Emoting with maps. *Primary Geography*, 67, 13–15.

North, W. (2009) Creating digital maps with primary pupils. *Primary Geography*, 70, 32.

Owens, P. (2014) Go wild, go explore, go digimap. *Primary Geography*, 83, 30.

Potter, C. and Scoffham, S. (2006) Emotional maps. *Primary Geography*, 60, 20–21.

Potts, R. (2011) Varmints! Storyboards and story maps. *Primary Geography*, 75, 9.

Pritchard, J. (2008) Worldmapper. *Primary Geographer*, 67, 30–33.

Schmeinck, D. (2007) Making a case for maps. *Primary Geographer*, 63, 36–38.

Scoffham, S. and Vujakovic, P. (2006) Edible maps. *Primary Geographer*, 60, 22–23.

Simpson, A. (2013) Maps, maps, maps. *Primary Geography*, 82, 12–13.

Tanner, J. (2012) How do you see it? *Primary Geography*, 78, 22–23.

Vujakovic, P. (2016) You are here. *Primary Geography*, 89, 8–9.

Webster, C. (2007) Projections and perceptions. *Primary Geographer*, 64, 24–25.

Atlas publishers

Three main publishers provide atlases for primary schools in Ireland:

- Folens: www.folens.ie/
- CJ Fallon: www.cjfallon.ie

- The Educational Company of Ireland: www.edco.ie/teachers/edco-primary-atlas-2012.3171.html.

Three main publishers in the UK provide atlases for Key Stages 1 and 2:

- Collins: www.collins.com/primary/geography
- Oxford University Press: www.oup.com/oxed/primary/geography
- Philips: www.octopusbooks.co.uk/philips-maps

Two *Barnaby Bear* atlases for Key Stage 1 use are available from the Geographical Association web shop.

References

Cicalò, E. (2017) Drawing and cognition in learning graphics and in graphic learning. *Multidisciplinary Digital Publishing Institute Proceedings*, 1(9), 1080.

Danos, X. (2014) *Graphicacy and culture: Refocusing on visual learning.* Shepshed, Leicestershire: Loughborough Design Press Limited.

Dolan, A.M. (2016) Place-based curriculum making: Devising a synthesis between primary geography and outdoor learning. *Journal of Adventure Education and Outdoor Learning*, 16(1), 49–62.

Forbus, K.D. and Ainsworth, S. (2017) Editors' introduction: Sketching and cognition. *Topics in Cognitive Science*, 9(4), 864–865.

Hart, R.A. (2013) *Children's participation: The theory and practice of involving young citizens in community development and environmental care.* Abingdon: Routledge.

Hennessy, B.G. and Joyce, P. (2013) *The once upon a time map book big book: Take a tour of six enchanted lands.* Westminster, MA: Candlewick Press.

Hennessy, B.G. and Madrid, E. (2012) *The scary places map book: Seven terrifying tours.* Somerville, MA: Candlewick Press.

Hope, G. (2008) *Thinking and learning through drawing in primary classrooms.* London: Sage.

Limerick Education Centre (2005) *Archaeology in the classroom – It's about time!* Limerick: Limerick Education Centre. www.itsabouttime.ie/

Mackintosh, M. (2010) Using photographs, sketches and diagrams. In Scoffham, S. ed., *Primary geography handbook.* Sheffield: Geographical Association, 121–133.

McOwan, P. and Olivotto, C. (2015) *How using mysteries supports science teaching* (2nd ed). Sheffield: TEMI – Teaching Enquiry with Mysteries Incorporated. http://projecttemi.eu/wp-content/themes/temi/pdf/Temi_teaching_guidebook.pdf.

Milne, A.A. and Shepard, E.H. (1957) *The world of Pooh: The complete Winnie the Pooh and the House at Pooh Corner.* New York: Dutton.

Proust, M. (2003) *In search of lost time: The prisoner and the fugitive.* London: Penguin UK. Originally published as *A la recherche du temps perdu.* Ed. Pierre Clarac and André Ferré. 3 vols. Paris: Gallimard, Pléiade, 1954.

Russell, K. (2008) A to Z of China: An alphabet journey through China. *Primary Geography*, 66(2), 26–27. Summer.

So, S. (2006) *C is for China.* London: Frances Lincoln Books.

Whittle, J. (2006) Journey sticks and affective mapping. *Primary Geography*, 59(1), 11–13. Spring.

Wiegand, P. (2006) *Learning and teaching with maps.* London: Routledge.

7 Teaching powerful geography through the arts

Introduction

The role of arts education in shaping the competences for young people for life in the 21st century has been widely recognised at European level (European Commission, 2009: 3). The term 'arts education' in primary schools generally refers to music, dance, drama and visual art. Arts practices have been part of the nature and expression of geographical knowledge for centuries. Original rock paintings simultaneously illustrate aspects of art and cartography. How do artists engage with 21st-century geography, with a world of incredible beauty alongside environmental degradation, with a world of technological advancement coupled with increased inequality, and with a world re-engaging with borders and exclusion? Art provides us with new perspectives, through which we can understand and challenge local and global complexities. More importantly, the arts have the capacity to offer hope and inspiration to children at a time when messages of hopelessness, doom and gloom are widespread. This chapter focuses on the potential of the arts for teaching powerful primary geography.

The chapter sets out to:

- explore the concepts of teaching geography for creativity and teaching geography creatively and how these can be assisted through arts education;
- examine how art can help children to engage with their local area through developing a curiosity and respect for its unique natural and human features;
- illustrate how geographical issues can be an inspiration for work in art, drama, sculpture and architecture;
- show how complex issues such as climate change can be explored through the arts;
- showcase innovative case studies from primary schools such as urban knitting, chainsaw art and landscape boxes; and
- demonstrate how artists and geographers can work together.

Creativity: Experiencing the world from a new perspective

Creativity is a complex and multifaceted concept not easily defined or understood. It is strongly associated with imagination, the emotions, play and new ways of thinking and seeing. According to the geographer Scoffham (2017: 3) creativity has a 'a number of dimensions ranging from the cognitive, the social and the emotional'. For Csikzentmihalyi (1997: 1) creativity is 'a central source of meaning in our lives'. Barnes (2011: 10) defines creativity as 'the ability in all humans imaginatively or practically to put two or more ideas together to make a valued new idea'. Craft describes the concept of 'possibility thinking' driven by children's questioning and imagination. With a focus on 'what if' and 'as if' questions, Craft (2002) argues that possibility thinking is at the heart of creativity. It involves a move from asking 'What is this and what does it do?' to 'What can I do with this?', particularly in relation to naming, refocusing and solving a problem. Her concept of 'possibility broad' tasks: allow children to use and develop their creativity in greater depth; push the boundaries of their knowledge and skills; provide them with the freedom to make mistakes and the opportunity to learn from them; and the space to reflect on their learning.

Scoffham makes a very important distinction between teaching creatively focusing on the teacher and teaching for creativity focusing on the learner. Teaching for creativity involves more control for learners, collaborative and cooperative approaches, enquiry questions and an emphasis on making connections. However, even though thinking creatively is part of the 21st-century toolkit, opportunities for developing creativity in the classroom are constrained due to limited encouragement, scripted instruction materials such as textbooks, increased accountability measures and high-stakes testing (Cremin, 2015).

There is tremendous scope for teaching geography creatively and for creative approaches to teaching geography including: the use of fun, games and playful approaches (Whyte, 2017; Witt, 2017); teaching geography though picture books (Dolan, 2017); innovative approaches to map work (Catling, 2017; Mackintosh, 2017; Pickering, 2017); the use of biscuits and cakes for understanding landscapes (Whitburn, 2017); local studies (Barlow, 2017; Pike, 2017); geographical opportunities provided by other curricular areas (Whittle, 2017; Tanner, 2017; Kelly, 2017); and education for sustainability (Owens, 2017).

While creativity in teaching and learning is not the preserve of the arts or arts education, there is nevertheless a special role for the arts within the exploration of geographical issues in the classroom. Research in neuroscience suggests that art enhances brain function and cognitive ability (Sousa, 2006). Howard Gardner's (2008, 2011) theory of multiple intelligences recognises the value of visual arts, music, dance and drama. Through these art forms children can express themselves but also find the tools to construct meaning and learn almost any subject effectively. This is especially true when the arts are integrated throughout every curricular area including geography. The arts provide an alternative means of communication, triggering emotional responses and naturally helping us learn. The value of creative approaches to learning has been supported by several educationalists including Froebel, Dewey and Vygotsky and more recently by Scoffham (2017), Pickering (2017) and Barnes (2018).

Geography has a unique function in that it serves as a bridge between the humanities and sciences. The arts have the capacity to leverage deep engagement and links with the natural environment. Offering a unique non-cognitive way of experiencing the world, art can reach the symbolic, sensory, perpetual, emotional and creative levels of human beings (Van Boeckel, 2009). Through art we can see the world afresh. According to the Finnish artist Osmo Rauhala (2003: 24), art is 'one of our antennae stretched out to sense the world'. Through art we can find new ways of existing and understanding our existence. Rauhala argues that art makes us predisposed to new information which may not be accessible to us in the traditional format of language. Through arts-based approaches, children can record and express their geographical understandings in several ways: writing poetry, drawing, painting, performing dance, music and digital film making. Art, through engaging the senses, can be a unique catalyst in developing a 'sense of wonder' about the local environment. By being flexible and spontaneous, powerful primary geography allows teachers to respond imaginatively to children's questions and interests, ultimately enhancing their creativity and their ability to think geographically.

Eco art

Biophilia, the love of the living world, can be developed in children when they have positive experiences in their local natural settings. Bringing children outside to explore with their senses, to observe deeply and to create art in their local natural or built environment develops a curiosity about and a respect for nature. Eco art also refers to environmental, land or ecological art. It integrates knowledge, skills, values and pedagogy from the visual arts, art education and environmental education as a means of developing awareness of and engagement with environmental concepts and issues such as place, interdependence, systems-thinking, biodiversity and conservation (Inwood, 2013). Eco art education embraces many of the 21st-century competencies. It teaches children to think creatively and to work collaboratively to solve problems with creative resolutions. More recently, environmental activism has stimulated the arts across the globe, for instance Björk in Iceland, Robert Bateman in Canada, Ansel Adams in the USA, Andy Goldsworthy in the UK and Arthur Boyd in Australia. Arthur Boyd, for example, painted a series of paintings communicating his pain over the ecological damage to his beloved Shoalhaven River, in south-eastern New South Wales, Australia.

Landscapes have always inspired artists but the idea of using the landscape as the canvas and available materials as resources is relatively new. Artists such as Andy Goldsworthy and Richard Long create art directly from landscapes. The sculptor Richard Long engages in art through walks in the countryside which he records through photography and word posters. Long has undertaken epic walks all over the world. Working with simple forms of circles and lines, he photographs his world as a permanent record. Long connects us with the timeless elements of nature. His work includes photographs and maps which mark his routes with spirals, circles and straight lines. Heavily influenced by geography, his work highlights an artistic sense of place as he demonstrates an understanding of the subtle qualities of landscapes (Romey, 1987). On his maps, he records events encountered on the way as text: the names of rivers crossed or sightings of clouds and changes in wind direction are recorded by arrows. Geographers can learn much from him about perception. His work explores relationships between time, distance, geography, measurement and movement.

Andy Goldsworthy is a British sculptor who creates temporary installations out of sticks and stones, twigs, leaves, snow, ice, petals, bark, rock, clay, feathers and any other material which is available at any given time. Resulting in ephemeral pieces which eventually disintegrate, he photographs each piece once completed. His projects are inspired by and interact with the natural landscape. His sculptures are born and die in line with natural cycles, their existence preserved only through vivid photographs and drawings. All of his pieces are designed to disappear as nature takes its course: ice melts, wind blows and rain falls, factors that shape how viewers experience Goldsworthy's constructions over the course of their temporary life spans. The final results are organised, colourfully radiating leaves, spiralling sticks, and mounds of thin rocks that convey the beauty of the natural environment in creative works of art. His outdoor sculptures represent sympathetic contact with the outdoor world. His artistic process is a poetic translation of material, place and the passage of time as it relates to seasons. Forces of nature including sunlight, sedimentation, tides, erosion, extremes of heat and cold, and plant growth and decay, can be experienced through his art. He has several beautiful publications (Goldsworthy, 2017) and any one of them will greatly enhance children's art practice.

Children love engaging in this kind of artwork. While its form is temporary in real life, its essence can be captured for ever on film. Student teachers find this artwork therapeutic,

184 *Teaching powerful geography through the arts*

Figure 7.1 Example of land art created by student teachers

creative and surprisingly affirming (Figure 7.1). By engaging in this kind of artwork children have a unique opportunity to spend extended periods of time in their own place, experiencing their place in new ways. This experience of their local place is geographic in nature as children learn to notice essential elements of their place which may have been heretofore unnoticed, unexamined or unadmired.

Artists have always been inspired by the world around them. They have captured perspectives of place through a variety of media including collage, drawings, maps, paintings, prints, models, ceramics, photographs, sculptures and digital images. Through a variety of art case studies, Barlow and Brook (2009) illustrate how children can explore their local areas with new eyes, and examine their values towards their communities in a way which promotes creativity and confidence.

Working through art provides children and student teachers with a unique opportunity to engage in problem-solving. Art provides space for children to solve problems without any restrictions. In their geography elective, student teachers choose an environmental issue and created an art installation which represented their geographical research, posed challenging questions and offered solutions. Their work showcased degrees of innovation, creativity and a level of problem-solving which impressed all those who had the privilege of attending the geographical and environmental exhibition '*Geography of our times*' (Figure 7.2).

Case study 7.1 Place-based learning: An artistic response to our local area

Cleggan National School is a small two-teacher school located in the Connemara National Park. Situated in County Galway in the west of Ireland, the national park is a stunning location which attracts tourists all year around. With scenic mountains, bogland, heaths,

Figure 7.2 'Geography of our times': Examples of art installations which address geographical and environmental themes

grasslands and woodlands, the park provides children with a treasure trove of wonderful geography on their doorstep. The park is home to the famous Connemara ponies, red deer and an enormous variety of bird life. Using their geographical location, the teachers adopt a cross-curricular place-based approach in their teaching. The following is a reflection from a teacher based on her use of the geographical locality for teaching:

> The children and I went for a walk in the national park. During the walk the children were shown how to identify the names of trees using leaves and a special tree identification sheet. Each child was asked to sketch their own drawing. Each child had their own viewfinder, pencil, and tree identification sheet. There were four stops along the route. At the first stop we looked at art in nature. The children were guided to a tree that had been felled by 'tunnelling' insects and we saw how it fell in a peculiar fashion, still being held up by two other trees. The children were asked to keep their 'creative eyes' open and to look out for any other such displays of 'art in nature' on our walk including autumn leaves or a nature scene.

At the next stop we looked at the waterfall and the ancient powerhouse located beside it. The children learnt that this little powerhouse first brought electricity to their local village nearly 100 years ago. We spent some time looking at and listening to the force of the waterfall as there had been heavy rainfall in the days prior to our walk.

At the third stop we looked at a sycamore tree that had an ash sapling growing out of a hollow in the trunk of it. We discussed how this may have happened and we highlighted how the ash tree is used for making furniture. Ash is also used in the production of a hurley (or camán) the wooden stick used in the national game of hurling.

At the fourth stop we arrived at the gates of the park and we looked at the holly tree and the hawthorn tree in detail, the children identified each one and they learnt about Irish folklore associated with trees. We also discussed the history of the woods in the national park and the children learnt that the woods were named after a Quaker family that lived there in the late 1800s. This family imported and planted thousands of trees that are in the national park today.

In preparation for their next walk in the woods the children were shown examples of art from Andrew Goldsworthy, creating art from nature. Inspired by Goldsworthy's work, the children had an opportunity to create their own art pieces. Working as photographers, two children documented the process. Some of the children made their pieces whilst on the Ellis Woods trail but most of the children brought their materials back to the playground to make their piece there. I was able to observe their work discuss it with them as it was being made and I was able to circulate around the playground to monitor their progress. The children were intensely focused on their work of art, paying attention to minute details e.g. getting the right colour leaf or the exact shape they wanted. The discussions I had with these children about their places were rich, interesting and meaningful based on many curricular subjects but mostly they were child-led. The children talked me through their progress, we spoke about colour, shape, symmetry, materials and nature in general.

At the end of our time in the park the class teacher and I allowed the children to have five minutes play time in the playground, some children chose this time to continue working on their piece while the others played. A few minutes later one of the children shouted at us to come over to him. As he was hanging upside down on the monkey bars he spotted something out of the corner of his eye … a huge dragonfly had landed on the wooden part of the monkey bars. We gathered the children around to see it and I pointed out the extraordinary colours this dragonfly had green and yellow on her head and the most beautiful blues on her long body. The children were absolutely amazed at this dragonfly. They thought it was false and that we had planted her there. This was a wonderful conclusion to our walk and it demonstrated the opportunities for unexpected and experiential learning outside the classroom.

After creating their art pieces outside the children sketched and photographed their work (Figure 7.3). Inside the classroom they began the next stage of the art project. Each child in the senior classes created a tile from clay using the design from their original piece (Figure 7.4). The completed tile montage is included in Section 8 of the colour plates. The pieces were painted and decorated. Children from the junior classes designed miniature ladybirds. Each piece was carefully assembled on a piece of wood which was beautifully framed. The final collaborative tile montage is proudly displayed in the school. The children succeeded in creating a striking piece of art based on inspiration from their local area.

Figure 7.3 Artwork based on the national park

Figure 7.4 Sample of the children's art work

The national park also hosts part of the Letterfrack Poetry Trail. Nine of Ireland's best poets were commissioned to write a poem celebrating the importance of place. Poets on the trail are Paula Meehan, Louis de Paor, Rita Ann Higgins, Theo Dorgan, Eva Bourke, Moya Cannon, Michael Gorman, Joan McBreen and Mary O'Malley. The poems are displayed on plaques in the locality. The President of Ireland Michael D. Higgins, also a poet, was invited to share one of his poems as part of this trail. The President's poem is in a secret location in the park and part of the mystery and magic around the trail is that participants have to find his unmarked poem using the clues given. The children have created their own poetry trails inspired by this place.

Architecture and drama: An unlikely alliance

Children have strong opinions about the quality of their local environments and geographical art facilitates them to articulate their likes and dislikes and ideas for change. Working as urban detectives, children can think geographically about their local spaces, draw buildings and design ideas for shaping the future of their areas. They can observe the design of buildings; the contribution of heritage; how transportation works; urban flora and fauna; processes such as the movement of products to and from shops; and the contribution of citizens and communities to local areas. Through geography children learn to pay attention to the built environment, look for details and understand that they are part of an amazing urban ecosystem. By working with practical materials children can represent and develop their ideas about their local area. Through drawings, plans and models, these ideas can be recorded and presented to a wider audience. Through an alliance of architecture and drama children can experience the opportunities, challenges and dilemmas facing local planners and architects.

Architecture is both an art form and a scientific endeavour, straddling most aspects of human thought and experience. Architectural design aims 'to bring an aesthetic harmony to the environment through creativity' (Gozen and Acer, 2012: 2225). Depending on its context, architecture can refer to the art of designing buildings, structures and outdoor spaces; a plan for organising space; buildings including skyscrapers, significant landmarks and schools; design for a city, town park or landscape; aesthetic engineering; or a carefully designed object such as a piece of furniture. Architects have been inspired by poetry, literature, geography, paintings, philosophy, scientific theories, psychology, religion, natural environment and mathematics. It is a way of viewing the world and shaping the world (along with geometries, tricks of light and, of course, plans). Our desire to build, both to provide shelter and to celebrate who we are and what we dream of, is innate (Burns and Dolan, 2019).

The Ancient Greeks were one of the first peoples to discover a way to harness the beautiful asymmetry found in plants, animals, insects and other natural structures. The Golden ratio/proportion or the Fibonacci sequence is a naturally occurring sequence of numbers that can be found everywhere in nature, from the number of leaves on a tree to the spiral shape of a seashell. It can also be found in famous works of art and architecture and even in our own faces.

Considered to be the most visually pleasing of all rectangles the Golden rectangle's sides are in proportion to each other. Golden rectangles can be found in the shape of windows, cards, ancient buildings and modern skyscrapers. The Ancient Greeks considered the Golden rectangle to be the most aesthetically pleasing of all rectangular shapes and it has been used many times in Greek buildings, including the Parthenon. Golden rectangles can be found in the shape, design and arrangement of windows.

The centres of most Irish country towns and cities were built during the 18th and 19th centuries and classically inspired architecture was the common style of the day. It can be recognised best by the use of the Golden rectangle. This shape was employed in ancient architecture, revived during the Renaissance and became popular as Neo-Classicism spread across Europe and on into Ireland from the 1660s. Georgian buildings make up much of the centre of Dublin, and during the 18th century this fashionable style of the day was disseminated throughout the country by way of architectural design books. While local tradesmen and masons used this architectural design, it was not always strictly applied, which leaves us with an idiosyncratic variation in many Irish towns.

Children are naturally interested in architecture. Give them a stack of paper and ask them to draw a house, or ask them to build a sandcastle, and they will be very happy. They enjoy making houses out of cardboard boxes, twigs, leaves, mud or stones. Children love to build structures around them which give them a sense of home, security and shelter. They have a sense of the rooms they inhabit and this can be seen when they rearrange bedrooms, playhouses or play areas. It can also be seen through the construction of replica houses with Lego and other construction material.

Like geography, space and place are two key concepts for architecture. We live in spaces created by the position of walls and borders. Even when we are outdoors we move about in spaces defined by buildings, walls, fences, roads and hedges. It is important for children to see that they can influence the quality of their local built environment. Urban spaces are constantly changing. Architecture involves an intuitive approach to creating the settings for our lived experiences. As citizens we can shape our local areas. Local environments can be changed and improved. It is important for children to learn not to take things for granted and to question why we do things the way we do.

Through geography and working as architects, children can observe, understand and enjoy the built environment. The local built environment provides an ideal resource for children to understand the design elements of public spaces and of private dwellings. Children are brilliant at noticing both positive and negative elements in design. Design generates a better understanding of our built environment and it helps us to imagine our local environment for future generations. Working as architects facilitates the development of 21st-century skills, mentioned earlier in this book, including higher-order thinking, problem-solving, creative thinking, visual thinking, group interaction and communication skills (Gozen and Acer, 2012).

Drama is an art form and also a powerful pedagogical tool. In Ireland the importance of drama is formally recognised in the primary curriculum. The emphasis is on 'exploring and making drama' as a means of investigating feelings, knowledge and ideas, and in examining relationships between people, in a real, historical or imagined context. From a geographical perspective drama presents children with an alternative route to understanding the world and thinking about their place in the world. Drama can bring abstract geographical concepts such as town planning to life in an engaging way for children.

The natural ability of young people to play provides drama with wide-ranging resources to facilitate learning. Drama in education gives children the tools to express themselves and imagine different worlds. The emphasis is on the process of creating drama rather than performance; exploring life through the creation of fiction rather than developing a piece of work for public display. The educational outcomes stressed by the Irish Primary Drama curriculum include the development of social, personal and drama skills (DES/NCCA, 2019). The way in which the child acquires knowledge through drama is as important as the knowledge itself. Drama allows children to collaborate and to be active in their own learning in line with constructivist approaches to teaching and learning. The planning activity in Case Study 7.2 illustrates the power of drama to facilitate real-life geographical discussions with children.

Drama in education employs theatre conventions such as still image, thought tracking and conscience alley to allow children to walk in the footsteps of others and to

190 Teaching powerful geography through the arts

explore values, emotions, ideas and decisions from different perspectives. Improvised role play provides children with an opportunity to create solutions facing different dilemmas (Heathcote and Bolton, 1994; O'Neill, 1995). By activating children's agency, the process of problem-solving gives children a voice. It develops their confidence and resilience. The strategy of teacher in role allows the teacher to participate in the improvisation, seeking clarification, asking questions or acting as devil's advocate. Once an improvisation is over children have the opportunity to critically reflect on the process including actions, responses and outcomes. This time for reflection allows the children to look back and critically reflect on their characters' actions and responses during the drama activities.

Case study 7.2 Planning through time

Enquiry question: How did building design from Ancient Greece find its way into the Main Street of your town?

Imagine if you could walk through your town or village and pick up any building and move it to a more appropriate site on the basis of a well-argued rationale. Children from Claregalway Educate Together NS (9–10 years) had an opportunity to do this through an artistic installation based on 3-D models of buildings in a 'typical Irish town' (Figure 7.5). A number of artists created an installation based on the evolving nature of Irish towns from the 18th century to modern times. This installation consisted of a miniature town with individual units which could be built and rebuilt by children (Figure 7.5). The exhibition highlighted classically inspired architecture found in typical Irish 19th-century main streets. Artists transformed plain cardboard boxes into brightly colourful buildings, with chimneys, roof angles and contemporary signage. Using these miniature cardboard buildings, children had the opportunity to design, develop and construct their own Irish town, while tracking the developments of different architectural periods. These model buildings formed part of an interactive installation designed to explore the expansion of Irish towns, historical building design and town planning. Children were invited to engage with the installation and undertake a variety of planning activities to enhance their understanding of the evolution of Irish urban landscape.

Figure 7.5 Planning Through Time Interactive Installation Project

Source: developed by Alison Mac Cormaic with creative advisers Dr Fiona McDonagh and Lali Morris

Curriculum-based art activities were designed to enhance children's understanding of the evolution of Irish urban landscapes. The installation explored the expansion of Irish towns and how they evolve organically as architecture changes through time. Through this work, children tracked the developments of different architectural periods and artistically explored heritage and architecture. Facilitated by the artists on site, children negotiated planning, development and the making of communities and made demolition decisions crucial to the development of their growing town.

Following this drama workshop, the children from Claregalway Educate Together NS looked at some of the issues raised during the workshop in the context of their local area.

Claregalway Place Notes

Claregalway is situated 10km north of Galway city in the west of Ireland. The town was built along the banks of the river Clare, hence its name. In Irish *Baile Chláir na Gaillimhe* means 'town on the Clare, in Galway'. While the village has an agricultural rural hinterland it is situated in Galway's commuter belt. Until recently the village sat at the junction of the busy N17 and N18 national primary routes with over 27,000 vehicles formerly passing through the village every day. Now that the M17 motorway has bypassed the village, traffic congestion remains an issue. Claregalway has been seriously flooded on several occasions. When the Clare River burst its banks, Claregalway village was split in two, and local farmers provided a temporary ford with tractor transport for several days. It is widely believed that overdevelopment including building on local floodplains has contributed to the rising water levels in the Clare River during the heavy rains. While millions have been spent on the Clare River Flood Relief Scheme, local people are still concerned about the possibility of future planning. The big issue for the children in Claregalway NS is their school. The children still go to school in a temporary building as they await the development of a new school.

The children were divided into six groups: Historians, Environmentalists, Children's Interest, Employment, Housing and Services. Each group was taken on a walk of their local area and asked to consider the following questions.

Historians

What are the historical sites in Claregalway?
Where are they situated?
Are they important?
Why should they be preserved?
What value have they for the people of Claregalway?
Why is history important for Claregalway?

Environmentalists

What aspects of nature are in Claregalway?
Why are they important?
How could Claregalway become a green village?

Children's Interest group

What facilities are in Claregalway for children?
What facilities do you use, e.g. education facilities, playgrounds, libraries?
What children's facilities would you like to see in Claregalway in the future?
Where do people work in Claregalway?
What kinds of jobs are in Claregalway?
What kinds of jobs will be in Claregalway in the future?

Housing group

Where do people live in Claregalway?
What kind of houses do they live in?
Where should people live in the future?

Services group

What services are currently offered in Claregalway?
What services should we think about providing in the future?

During their walk each group looked for geographical evidence in the town which addressed their theme (Figure 7.6).

The children took photographs and made notes and on their return to the classroom prepared their presentation for the class (Figure 7.7).

After the presentations the children returned to their groups and completed the following reflection exercise:

1 What did you learn?
2 What makes Claregalway special?
3 What issues are facing Claregalway today?
4 What will Claregalway be like when you are 40 years of age?

Figure 7.6 Children researching local issues in Claregalway through field work

Figure 7.7 Children presenting their research in class

What did you learn?

There are good and bad things about decisions.
Important not to build on marsh lands.
Sometimes in order to build things up you have to destroy land.
Start with a small thing and build from there.
Build over old towns but leave some historical things.
You can't always have what you want, you have to make choices.
Different buildings have different uses.
Sometimes you have to destroy things to create new things.

What makes Claregalway (our place) special?

Its unique history: Claregalway has its own abbey and castle.
It was a small community which became very built-up due to its proximity to Galway.
We have everything here. It's like a mini city, it takes too long to travel to Galway.
Personality of the people.
The community centre: you can play sports there and we have a stage for drama.
The geography of the village including the river Clare, the housing estates, the Nine Arches.

What issues are facing Claregalway today?

Clare River flooding.
Litter in our village.
Pollution – lots of cars/fumes causing global warming.
Pollution of the river.
Lack of money – things that need to be built still aren't built, like our school, a bungalow project for the elderly.
Ghost estate – lots of unfinished houses without windows.

People want space to build more and more things.
Impact of motorway:

- Land taken from owners that they need for their cattle.
- Although you're getting places quicker it's not helping the environment.
- More difficult to get to school in the morning.
- Not safe for pedestrians.
- Fumes from cars affecting farmland and food.

What will Claregalway be like when you are 40 years of age?

Different architecture.
Bigger city.
This building will be a nursing home.
More skyscrapers.
Global warming – this place might be a bit hotter.
Less green land. Galway city might be out as far as Claregalway.
Professional city – there will be a lot more money here.
Claregalway could be smaller because of flooding.
A floating city, this will be a darker place.
There was a boom the year before I was born so a boom could happen again.

The final activity involved an exercise for children to apply all of the learning from earlier activities. Once again children worked in their specialist groups. They were informed that the government had granted €2 million funding for Claregalway. A number of scenarios were presented to the groups for their consideration and each group had to select two options for funding. The scenarios presented were as follows:

1. Refurbish and develop Claregalway Castle as a significant tourist attraction for the town.
2. Money to prevent river Clare from flooding as there is no guarantee that the current measures will work.
3. New housing estate which will house 300 people. There is a severe shortage of housing in Claregalway.
4. Family fun complex with cinema and extensive facilities for children.
5. Factory which will create 300 jobs for Claregalway. People have to commute long distances for work.
6. New school building with state-of-the-art facilities including new library, computer room, music facilities, art room and sports fields.
7. New bypass of the village to alleviate the traffic congestion in Claregalway. The new motorway has still not solved the traffic problems in Claregalway.

Interestingly, the majority of children voted for a new school and funding for flood relief as these are the issues which have a direct impact on their lives. The transformative potential of this initiative is discussed later in this chapter and illustrated further (Table 7.1 and Figure 7.16).

Case study 7.3 The Bees' Big Day Out

Children from St John the Apostle NS, Knocknacarra participated in a very exciting whole-school community art project based on bees in their local environment. The project was inspired by a bronze beehive sculpture (Figure 7.8). Created by Mark Rode, the sculpture in the school's front garden was funded by the Department of Education and Skills under the Per Cent for Art Scheme.

The school community set out to create an original story and devise a corresponding script of soundbites which incorporated the voices of the children. The artistic collaboration generated *The Bees' Big Day Out*, an original and exciting new story which brought people on a unique adventure through a colourful and enchanted world of art installation and soundscapes. The design and layout for the installation came first while the framework for the soundscape evolved some weeks later.

Parents, teachers and children worked together to create the multi-sensory installation. Each child made their own special world bee. One parent coordinated this process including design, development, threading and hanging. Other parents learnt how to weave, then designed and constructed an amazing willow beehive for the exhibition. Another team was responsible for the creation of the skyscape (soft sky with 2-D and 3-D clouds).

One of the dads shared his musical expertise through the composition of the soundscape and provision of technical expertise. He was responsible for the recording and engineering side of the project and set up a studio in the school for all children to take part in the recording sessions adding their voices through song and soundbites to the story *The Bees' Big Day Out*. Another dad brought his wizardry as a lighting technician to the project to add a whole dimension of colour, movement and dynamism to the exhibition space. The artistic process was assisted and supported in all sorts of ways by the input of the Bee-dazzling children, their teachers, school staff, parents and friends from the arts, all of which led to the creation of a whole-school community arts project.

The soundscape developed into an 18-minute collage of music, nature sounds, song and dramatisation. This became the avenue for narrating the magic story of the bees from all over the world on their journey from the hive, through the vibrant world of art installation, to find the greatest flowers on Earth and to ultimately locate 'The Nectar and Pollen Paradise'.

Figure 7.8 Bronze beehive sculpture which inspired The Bees' Big Day Out

This artistic collaboration culminated in a wonderful and enchanting new world of story, sculpture, textile, art, music, colour, song, voices and magic, magic, magic for all. The school community succeeded in creating something very special, a world that mirrors and reflects the imagination, colourfulness, playfulness, fun and intelligence that exists so naturally within the precious world of the child, a special and beautiful world now known to all as *The Bees' Big Day Out*, which exists only high up in a mystical place, known to some as 'The Bee Room' in St John the Apostle NS Knocknacarra. Some images are included in the central photocollage of this book. The geographical and cross-curricular learning from this project was immense. Bees are essential because they are very important pollinators of plants. They help pollinate both crops and native plants, hence their economic and ecological importance. In Ireland there are 100 species of bee. However, 50% are in decline with a third threatened with extinction. A combination of habitat loss, intensification of agriculture, hunger due to a decline in wildflowers, disease, poisoning by pesticides, climate change and imported species has caused bee population decline in Ireland. The UN has designated 20 May as World Bee Day to highlight the importance of honeybees as well as all the other bee species, and their role in ensuring global food supply. Children from St John the Apostle NS, Knocknacarra are deeply aware of the local and global importance of bees and more importantly they are happy to work as local ambassadors for bees (Sections 4 and 5 of the colour plates).

Case study 7.4 Chainsaw art

Chainsaw art is the ultimate form of recycling; it involves creating fine art from pieces of wood. According to local Limerick wood sculptor Will Fogarty, 'a chainsaw carving is a great way to add interest and beauty using the most natural material in its own habitat'. Following a tree fall after a storm, Fogarty created three new wooden sculptures for the People's Park in Limerick. He carved three dogs: a Springer Spaniel, a Scottish Terrier and a Jack Russell. The inspiration came to the wood sculptor when he visited the park and noticed all the people who walk their dogs there.

Will began carving a few years ago making walking sticks and staffs, which are made from hazel collected in local forests. While he still makes these he has moved into chainsaw carving and found his passion. All of Will's work is done on wood that has been felled by nature or has been cut down in a sustainable way. Will gets a huge feeling of joy and satisfaction as he witnesses the beauty emerging from the wood, and he knows that each piece of wood will be enjoyed for years.

Scoil Íde, Corbally in Limerick have also benefitted from Will's artistic talents. One Winter a beautiful lime tree was badly damaged during a storm. As a result, the tree had to be cut down for safety reasons. Having seen some 'chainsaw art' in the People's Park from previous storm remains, the school contacted Will to see what he could do with the stump. While the school community deeply regretted the loss of their beautiful lime tree, they couldn't have envisaged the transformation of the stump to the beautiful piece of art it has become.

Will has worked on the theme of creatures local to the area. He has carved an owl, heron, salmon, kestrel, badger and squirrel from the stump at the time of writing (Figure 7.9). The project has garnered huge interest both locally and through social media where the school's Facebook page had a 'reach' of over 10,000 people. The tree sculpture is expected to last over 30 years and it is hoped that many children and parents will enjoy it. This

Figure 7.9 Scoil Íde's magnificent tree sculpture

piece of art is now a prominent part of the school geography and is enjoyed by parents, children and teachers. Symbolically, it inspires the children to think about their local area from ecological, sustainable and aesthetic perspectives.

Teaching complex issues such as climate change through the arts

Art has the capacity to communicate complex geographical issues such as climate change, deforestation and urban decline. By using metaphors, analogies or narratives, art can make abstract concepts more accessible, giving the viewer a personal experience (Roosen et al., 2018). This is especially important as many people still see climate change as an abstract issue that poses no direct threat. In Leonardo DiCaprio's documentary *Before the Flood*, he begins with a story about a painting from his childhood. The 500-year-old *Garden of Earthly Delights*, painted by Hieronymus Bosch, hung on a wall in his bedroom. Broken down into three scenes, each on a different panel, the painting appears to portray the way humans can be misled by life's temptations, and how that can eventually lead from an Eden-like existence to a hellish nightmare. DiCaprio says he stared at the painting as a child until it told a story: the first panel is full of animals and spectacular scenery; the second panel is where the deadly sins start to 'infuse their way into the painting' along with overpopulation and excess; and the last 'nightmarish' panel is a 'twisted, decayed, burnt landscape' and a destroyed paradise. Art historians often decode the painting as a clear premonition of the risks of temptation and the misuse of nature. The Bosch painting provides a powerful parallel to the narrative of DiCaprio's documentary, feeding the imagination as the documentary itself provides the scientific analysis. While the science of climate change communication is essential to engage people's minds, the art of engaging people's imaginations may be just as important (as discussed in Case study 7.5).

Case study 7.5 Esker Educate Together National School, Lucan

Esker Educate Together National School has a magnificent sculpture located in its school garden which reminds the school community of the challenges posed by climate change (Figure 7.10). Titled *The Book of Climate Bells*, the sculpture, which was designed by Vivienne Roche, was officially unveiled by President Michael D. Higgins at a very special ceremony on International Children's Day.

The sculpture is deeply symbolic. Vivienne Roche's Climate Bells were inspired by the address President Higgins gave at a previous climate change meeting in which he said that 'each citizen will be just as decisive in reaching the ambitious agreement mankind needs as heads of state'. Smaller versions of the Climate Bells have been designed as state gifts for world leaders including Pope Francis, President Santos of Colombia and President Kuczynski of Peru. The bells signify the community effort that President Higgins believes is needed to limit climate change.

The launch of the Book of Climate Bells was a special event in the school history as described here by the school principal:

> Children from the Green School/Global Citizenship Committee made a presentation to the President about climate change. The President was told about how children can make a difference to our planet and how we should try to reduce our ecological footprint. One of the young poets recited an original piece about how we could ensure our successful future by minding our earth and changing our wasteful habits. A short walk, through a very lively and excited guard of honour of senior infants, led to the gazebo, where artist Vivienne Roche discussed the art installations in the garden. The President opened *The Book of Climate Bells* and two sixth class children and one of the teachers rang the school bell three times to signify a warning bell, a call to action, and the power of individual and group action in looking after our environment. The party then moved to the Sunbell Garden, designed by Vivienne Roche. This outdoor classroom is made of Corten steel with a Perspex sun and old ship's bronze bell. When the sun shines at midday, it throws a beautiful light through the window on to the ground.
>
> Senior infants told the story of Eddie the Penguin, who wanted to save the earth. They were joined in song by the choir. The choir sang about the future being in their 'Own two hands'; that they can effect change by taking responsibility for our planet. The President was presented with a specially commissioned calligraphy version of the 'Proclamation for a New Generation', which was written for the centenary of the Easter Rising 1916. The choir finished with their school song, 'What a Wonderful World'.

Figure 7.10 Student teachers visiting the Book of Climate Bells

Bells are symbols of communication, communicating ending points, celebratory points, initiating points, tolling at times of practically all the experiences of human beings. They mark the rhythms of school and community lives reminding people of possibilities and important responsibilities. This book opens and closes on different occasions reminding all at Esker Educate Together School that a whole new chapter is beginning, and it is up to the children to write the story. *The Book of Climate Bells* is designed to create and maintain a consciousness about climate change, an issue which is discussed in greater detail in Chapter 5.

Children and teachers from the school community invest considerable time in articulating their hopes and vision for the future. Through their articulation of a 'preferred future' they ascribe to Hicks' notion of futures education. During the national celebrations for the 100th anniversary of the 1916 Rising, the children wrote their own proclamation for a new generation. This proclamation was beautifully scripted and framed. It is now prominently displayed in the school building. A framed copy was also presented to the President of Ireland. The process of drafting the proclamation represents the children's ideas in terms of 'possibility thinking' (Craft, 2002) promoting their agency and confidence.

Proclamation for a New Generation (compiled by children from Esker Educate Together NS)

We stand here today to proclaim our hopes and wishes for the Ireland of our future, the Ireland we want for ourselves and for future generations. These wishes represent the vision expressed by the children of our school and it is our hope that this is the Ireland we will grow up to see.

The Ireland we hope for will be a place which is fair and free from fighting.
We hope for an Ireland where all people will have homes where they will feel safe, where we can all live united in peace and security,
where family means love, and different kinds of families are welcome,
where everyone will have the chance to get a great education and learn to love books, and where disagreements are solved in a fair and orderly way.
We hope for an Ireland where we will all look after the environment,

> where there will be more green places for children to play,
> where more trees will be planted and people will stop littering.
> We hope for an Ireland where there will be equal rights for all, regardless of skin colour or where you are from,
> where boys and girls enjoy equal respect and opportunities,
> where everyone has the right to follow their religion in peace,
> where people with disabilities are treated with dignity and fairness,
> where we will all have the right to make our own choices and have our own opinions.
> An Ireland where we will find a way to cure sickness,
> where we will see friendship and kindness,
> where everyone will learn to get along,
> where people love each other and nobody is lonely,
> and where there will be no more bad dreams for anyone.

Case study 7.6 Shaping spaces: Geographical art in the school grounds

School grounds and the school yard can reflect children's geographical ideas through art and design. On my first visit to Scoil Íde, Salthill, Galway I was inspired by the murals on the school walls. The principal proudly informed me that they were designed and painted by the school children themselves. The school yard of Scoil Íde features several murals which highlight geographical themes and events. Each year sixth class (12–13 years) decide which theme they would like to design. Working with leftover paint from home together with minimal assistance from a local artist, they spend three full days working on their project. The murals give the children a sense of ownership of the school grounds and it is with great pride that they regularly return with visitors to display their work of art. Their work lives on as an inspiration for future school generations of children.

Each year the children consider local events and the geography of their local area. First, the children brainstorm ideas for their design. Here are the ideas from one class:

1. Burren, sea and mythical creatures;
2. The Galway races;
3. Flora and fauna from the Burren;
4. Salthill beach;
5. Cinema and people watching a film;
6. Cultural panel featuring art, literature, music, drama and dance;
7. Salthill prom and park;
8. Galway Arts festival;
9. Collage of Galway life;
10. Streets of Galway;
11. Photocollage of important themes;
12. Our school.

After a vote the children decide to work on the theme of the Burren and mythical creatures. From Salthill the children have a magnificent view of Galway Bay and the distant Burren and in low light the views of the Burren are used to inspire the children to think about mythical life (Figure 7.11). More images are included in the colour plates (Section 2).

Figure 7.11 Schoolyard art

Case study 7.7 Urban knitting

Urban landscapes include public sculptures, monuments and buildings with a variety of architectural styles. Geographers engage with various forms of urban interventions which question and challenge the power dynamics, aesthetics and layout of urban spaces (Price, 2015). Informal art practices such as street art are used to question the existing environment. Urban knitting, a relatively new form of the street art genre, is used to highlight aspects of the local environment, to draw the viewer's gaze and to change the perspective of something which is perhaps taken for granted or remains 'unnoticed'. Also referred to as 'knit graffiti', 'yarn bombing' or 'guerrilla knitting', urban knitting as an art form is thought to have originated in the USA when knitters started wrapping their knitting around public property, covering street signs, fire hydrants and trees with their colourful, woolly creations.

Geographers have long recognised the significance of play and playful creativity in all environments both urban and rural (Price, 2015). Urban knitting involves attaching a hand knitted or crocheted item to a street fixture, public monument or part of a public building. It can be a complex piece large enough to cover a public monument or it can be small and discreet such as a covering for a street sign. Like many forms of street art, urban knitting is linked to social action, the local environment and creative expression.

Eyre Square is located at the heart of Galway city. Named after the Eyre family discussed in Chapter 2, the square is one of the famous landmarks associated with Galway. The square has a rich history dating back to medieval times when markets took place on the green in front of the town gates. One of the iconic symbols of Galway is the Quincentennial Fountain (1984), a representation of the Galway Hooker, a traditional fishing boat unique to Galway. The fountain centrepiece consists of a copper-coloured representation of the sails of the Galway Hooker. A recent addition to Galway's artistic and architectural heritage, it manages to embrace Galway's traditions in a contemporary fashion.

Urban knitting involves people covering public monuments with knitting. The Sail sculpture in Eyre Square was yarn bombed by children from Tirellan Heights National School. A teacher at the school – Karla Bodeker – saw a book on yarn bombing and asked the children at the school if they would like to make an installation. Once they received permission from Galway City Council the children started to knit. The aim was to knit enough squares to decorate the sails of the fountain in Eyre Square (Figure 7.12).

202 *Teaching powerful geography through the arts*

Figure 7.12 The Eyre Square Monument yarn bombed by children from Tirellan Heights National School

Source: image supplied by Ailish Farrelly

At the beginning of the project most of the children could not knit and were taught by their grandparents, aunts, uncles and friends. The children knitted squares featuring every colour of the rainbow. The caretaker measured the sails and put sticky-tape on the hall floor the same size as the sails. When the knitted squares were completed, the children started to sew. The knitting took almost 5 months and 134 children participated in the project using 300 balls of wool to knit 2,086 squares.

Once the knitting was installed on the monument the Urban Knit Project was launched by the Lord Mayor. The entire school community was present. The project was featured on local and national media and the children were interviewed by several media outlets.

The children were interested in continuing their urban knitting project in other geographical locations. One of the famous landmarks in Galway, the bronze sculpture of Irish writer Oscar Wilde together with Estonian writer Eduard Vilde, was also yarn bombed by the children. For Christmas, the children knitted Santa suits for the two iconic figures. This Christmas knitting project inspired the children to study urban features in their locality. The Eduard Vilde statue by Estonian artist Tiiu Kirsipuu was presented to Galway when Estonia joined the EU in 2004. Tourists often pose for photographs beside this famous monument (Figure 7.13). The sculpture is a replica of the unusual streetscape image that sits outside the Wilde Irish Pub in Tartu, Estonia. The two Wildes were contemporary

Figure 7.13 The bronze sculpture of Irish writer Oscar Wilde together with Estonian writer Eduard Vilde, a famous landmark in Galway City

Irish and Estonian writers and this sculpture imagines their possible meeting in 1892. The people of Estonia have for some time stated their enthusiasm for making a copy of the sculpture and presenting it as a gift to Ireland.

Kent and Mackintosh (2015) demonstrate how works of art enable children to understand that different people see the world in a variety of ways. Art can also help children to develop their own views about their local area. Based on her ideas of bringing children to view a local sculpture, I brought a group of children to the sculptures of Oscar Wilde and Eduard Vilde. We explored the possible identity of the two men through questions such as: Who are these men? What are they doing? Why are they here? What do you think they might be talking about? What view do they have of life in Galway? Why did the artist create this sculpture? Back in school they created a still image of the sculpture. A thought-tracking technique was used to enable the children to talk in role as one of the characters in the sculpture. Kent and Mackintosh (2015) remind us about the importance of offering questions which promote geographical thinking, such as: What is the weather like? What do you think about your surroundings or view? How could this area be improved? Does their view of Galway represent a realistic perception? The children then had an opportunity to design and make their own sculpture. Every sculpture and monument has an important impact on the local physical and cultural environment. Initiatives such as urban knitting or art appreciation provide an opportunity for children to pause, ask questions, research and create a new chapter for their local geographical and historical monuments.

Case study 7.7 Scrapbooking

The act of putting together a place-based scrapbook is a multi-sensory experience. Scrapbooking is a creative exercise where the child shapes the process and the product. Some scrapbooks are more visual while others are annotated with detailed descriptions. The act of selecting, cutting and pasting images helps a child to prioritise, reflect and create a layout which is pleasing to him/her as author and to an audience with whom the book may be shared.

Scrapbooking provides a useful way of exploring children's personal geographies (Witt, 2010). Place-based scrapbooking helps children to see geography all around them and to

Figure 7.14 Example of geographical scrapbook recording

realise its significance in their daily lives. Children learn through their own observation and creating a scrapbook provides a motivation for looking carefully and recording accurately. It provides an opportunity to include their own interests and abilities in the learning process. Scrapbooks encourage thinking at the higher levels of Bloom's Taxonomy, specifically application and evaluation. The process of scrapbooking gives children a sense of ownership because they invest so much of themselves in the project.

Children from Carnmore NS were invited to complete scrapbooks based on their journeys during their summer holidays. This was a voluntary exercise. Scrapbook entries included photographs, drawings, images, maps, notes, memorabilia, newspaper clippings, tickets from matches and brochures from hotels (Figure 7.14). All kinds of written entries were encouraged including speech bubbles, poems, short stories, long pieces both handwritten and typed. The scrapbooks capture a sense of the places visited during the summer. They included holidays in Ireland and abroad, day trips to matches and family outings.

The following guiding questions were provided:

Where is this place? Map or description of location
What is this place like? Description of image (photography or drawing)
What does the child think about the place? Like? Dislike? Reason? Opinions?
Any interesting or unusual stories about this place?

The scrapbooks provided a personal record for the children of their geographical engagement with place. It was an enjoyable experience and a source of pride for the children. Parents commented on the usefulness of the exercise as it provided an opportunity to reflect on geographical experiences, it involved the whole family and it was fun. As a form of communication scrapbooks can be recognised as a form of literacy (Gee, 2012). The multimodal nature of scrapbooks facilitates creativity, flexibility and non-traditional presentation of geographical data. However, scrapbooking should be linked to clear geographical learning objectives in order to maximise geographical learning (Witt, 2010).

Case study 7.9 Landscape boxes

Parkinson's 'miniature landscapes' (2009) are a wonderful way for children to engage in researching natural and human elements of different places. Using a shoe box or similar-sized box, children can build a landscape. Children can create landscape boxes based on their own local area or, depending on the theme of their study, their landscape box may feature somewhere farther afield. Every part of the box can be used to illustrate features of the landscape including the lid and front panel of the box (Figure 7.15). Children can sculpt their boxes with images, photographs, paper mâché, recycled materials, miniature people and animals. Written descriptions can enhance geographical learning. Once children prepare their boxes they can present them to the class outlining reasons for choosing their place and describing features in the box.

Materials needed:

- Computers or tablets for research and access to photographic images;
- Shoe boxes or larger boxes such as pizza boxes if desired;
- Coloured sugar paper and/or construction paper;
- Glue/glue sticks/sticky-tape;

Figure 7.15 Landscape boxes

- Other possible materials include cardboard tubes, foam sheets and feathers;
- Maps;
- A landscape box planning sheet;
- Coloured paper and magazines;
- Stiff paper for inserting drawn images to make them stand vertically;
- Markers;
- Glue (hot glue sticks and a hot glue gun may be most useful);
- Tape;
- Scissors;
- Miscellaneous materials to create 'realistic' miniature versions of rocks, soil, plant life (clay, reindeer moss, gravel, paper, cardboard, paint).

Divide the children into groups and present each group with a detailed map of their area (preferably an Ordnance Survey map). Ask the children to locate their school building. Ask the children if they think of any local landmarks that may be on the map, such as an old church building or ruins, rivers, main roads or the town centre? Once the children have located some features ask them to think about the unique aspects of the environment where they live. Why is a particular feature there? How long has it been there? What does it tell us about our local area? Children can group all the features they have noted under the following headings: Place names, Religion, Archaeological remains, Education facilities, Buildings, Roads, Land and agriculture, Trade and industry, Social life, Transport, Public utilities, Law and order, and Military. Children can choose an aspect they would like to feature for their landscape box. To ensure maximum learning potential, the boxes can be displayed in the school or in a local library, or other public venue.

An interesting contribution to place-based learning, landscape boxes can also be designed in conjunction with local trails, as described in Chapter 3. They can also be used to illustrate famous landmarks in Ireland such as Newgrange, the Cliffs of Moher, Dún Aengus, Christ Church Cathedral or the Burren. Discuss with the class how such places are part of Ireland's national heritage. As well as the most famous national sites, every local area has its own heritage and history. International biomes and landmarks can also feature in landscape boxes.

Landscape box: Planning sheet

- Title of your box
- Background research about the landscape
- List the things you need to include in your box
- What everyday items might you use
- What other techniques might you use for the background, foreground and middle ground of your landscape
- What have you learnt from this exercise

The 'Landscape in a Box' project gives children the opportunity to demonstrate their understanding of physical, human and environmental geography by recreating a famous world landscape in a shoe box. As well as recreating these wonderful landscapes, children also prepare a one-minute speech which they deliver to the class. This speech highlights the physical, human and environmental aspects of landscape, possible changes this landscape could face in the future and ways of managing these changes. Boxes can also be exchanged between classes from different geographical locations.

The arts as a catalyst for transformation

Powerful primary geography has the potential to be highly creative. As a transformative practice, it provides a space for children to author their visions of a better world. Case studies in this chapter demonstrate how the arts provide a mechanism for children to feel, imagine, create, relate and express themselves. Through the arts children can articulate and create the world they want to live in. This involves the engagement of multiple intelligences and artistic media including theatre, music, visual art, photography and dance. Accessing the imagination allows children to see geographical concepts through a new lens, to see them other than the way they are. This is what artists and scientists do according to Eisner (2002: 199), 'they perceive what is, but imagine what might be, and then use their knowledge, their technical skills, and their sensibilities to pursue what they have imagined'. Children working as geographers can imagine 'preferred' and 'probable' futures and decide what needs to happen to achieve their preferred future (Hicks, 2014). Informed by children's questioning and imagination, Craft's (2002) concept of 'possibility thinking' promotes children's agency and confidence in their own ideas (Table 7.1).

The potential for geographical engagement with an arts initiative can only be achieved through sufficient time, appropriate enquiry-based activities, space for the children to reflect and teachers' competence in directing the learning initiatives. Taking the example from Case Study 7.2. the children from Claregalway Educate Together NS were fortunate to have the opportunity to participate in an arts-based town-planning exercise. But the geographical learning was maximised by sustained follow-up activities in the classroom and their local area over a four-week period. Table 7.1 and Figure 7.16 take this example of town planning and illustrate the potential for an artistic installation to facilitate powerful primary geography.

Table 7.1 Teaching geography powerfully through the arts

Maude's five types of powerful knowledge	21st-century competencies	Teaching geography powerfully through the arts
Knowledge that provides children with new ways of looking at the world	Knowledge building	Engage in town planning on the basis of sound geographical knowledge and skills
		Experience the dilemmas involved in town planning
		Engage in decision-making
		Recognising the implications of one's decisions
Knowledge that provides children with powerful ways to analyse, explain and understand the world	Critical thinking	Asking geographical questions, e.g. where is the best place for locating a school? What are the implications for demolishing a historical building?
		Involved in futures thinking, e.g. what do we want our place to be like in 30 years' time?
		Assessing the environmental and human impact of town planning

(Continued)

Table 7.1 (Continued)

Maude's five types of powerful knowledge	21st-century competencies	Teaching geography powerfully through the arts
Knowledge that gives children some power over their own knowledge	Real-world problem-solving	Identification of local issues, challenges and problems, e.g. local flooding
		Formulate solutions to problems, e.g. recommendation to ban building on flood plains
		Discuss and debate the merits and demerits of different solutions
		Assess the human and environmental impacts of each solution
Knowledge that enables young people to follow and participate in debates on significant local, national and global issues	Collaboration and teamwork	Articulating opinions and options within one group in relation to solutions for one local problem, e.g. flooding
		Considering all options
		Agreeing a group solution
		Articulating the group solution within and beyond the group
		Becoming a group ambassador
Knowledge of the world	Global competency	Understanding how Claregalway is affected by local, national and international events such as climate change, challenges caused by the boom and bust of economic cycles and the introduction a new motorway
		Making links with other schools, villages and places affected by similar issues, e.g. using Skype to share project work
		Sharing children's reflections, images and maps with a broader audience through social media, local newspaper or national competitions

Conclusion

Cross-curricular learning involving geography and art is a powerful way for developing 21st-century competencies including critical and creative thinking, collaborative learning and problem-solving. Art provides a unique way of understanding and responding to the world which is ultimately a geographical aspiration. Working with the arts involves a learning process which is experiential and open-ended. Perceiving art demands attention, and processing art requires parts of the brain not normally associated with language-based learning. The arts make the familiar unfamiliar or in the words of Williamson and Hart (2004) they make the ordinary extraordinary. They invite us to review and renew our understandings of everyday events and surroundings. The arts provoke us to think differently and to look at something with new eyes. This chapter demonstrates the power of the arts for making complex geographical ideas and themes accessible to children. Children's agency is

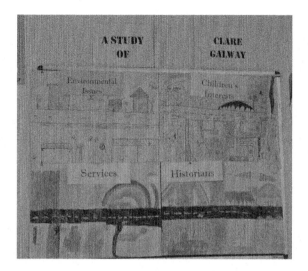

Figure 7.16 Articulation of children's place-based ideas through their maps

promoted through their responses, enhanced imagination and multi-layered ideas. Children develop an understanding of their role as citizens as they articulate their views and the values shaping these views. They develop knowledge of the importance of taking action and how actions can change the world for better or for worse for future generations. As citizens children have the right to have their ideas about the quality of local environments heard. Powerful primary geography provides a space for these ideas to be constructed, articulated and shared with a broader audience. Conversely, improved environments will only be achieved when the public become involved in decisions shaping their local spaces.

Exercise 7.1 Reflection for teachers

How can your local geographical area be an inspiration for artistic exploration?
Could the geographical concepts in your long-term and short-term plans be further explored through an artistic lens to enhance your geography teaching?
In the spirit of cross-curricular teaching how can the arts and geography be delivered in a way which enhances children's exploration of big geographical ideas discussed in Chapter 1?
How can the arts provide children with a new lens for exploring issues in local, national and international contexts?

Further resources

Barnes, J. (2018) *Applying cross-curricular approaches creatively*. London: Routledge.
Mackintosh, M. and Kent, G. (2014) *The everyday guide to primary geography: Art*. Sheffield: Geographical Association.
Pickering, S. (2017) *Teaching outdoors creatively*. London: Routledge.

Website

www.heritageinschools.ie

Articles published in the Geographical Association's journal *Primary Geography*

Barlow, C. and Brook, A. (2010) Geography and art part 1: Local area work 16–17. *Primary Geography*, 72(2), 16–17.
Clayton, J. (2016) Art and the locality. *Primary Geography*, Spring 30–32.
Horler, T., Mackintosh, M., Kavanagh, P. and Kent, G. (2014) The art of perceiving landscapes. *Primary Geographer*, 83, 8–10.
Kent, G. and Mackintosh, M. (2015) Art and geography: Children's own places. *Primary Geography*, Spring, 26–27.
Mobley, A. (2014) Earth as art. *Primary Geography*, Spring, 83(1), 6–7.
Witt, S. (2010) Geography and art part 2: Happy spaces, happy places – Exploring pupils' personal geographies using scrapbooking. *Primary Geography*, Summer, 72(2), 18–19.
Witt, S. and Sudbury, J. (2010) Geography and art part 3: A sense of place at Bishop's Waltham Junior School. *Primary Geography*, 72(2), 20–21.

References

Barlow, A. (2017) Geography and history in the local area. In Scoffham, S. ed., *Teaching geography creatively*, London: Routledge, 118–130.
Barlow, C. and Brook, A. (2009) Valuing my place: How can collaborative work between geography and art help make the usual become unusual? *Cross-curricular Approaches to Teaching and Learning*, 49–74.
Barnes, J. (2011) *Cross-curricular learning* (2nd ed.). London: Sage, 3–14.
Barnes, J. (2018) *Applying cross-curricular approaches creatively*. London: Routledge.
Burns, I. and Dolan, A.M. (2019) Exploring settlement though architecture and art. *Primary Geography*, 99(2), 14–15.
Catling, S. (2017) Mental maps: Learning about places around the world. In Scoffham, S. ed., *Teaching geography creatively*, London: Routledge, 58–75.
Craft, A. (2002) *Creativity in the early years: A lifewide foundation*. London: Continuum.
Cremin, T. (2015) Perspectives on creative pedagogy: Exploring challenges, possibilities and potential. *Education*, 353–359.
Csikzentmihalyi, M. (1997) *Creativity: Flow and the psychology of discovery and invention*. New York: Harper Collins.
DES/NCCA. (1999) *Primary School Curriculum Drama: Arts Education*. Dublin: The Stationery Office.
Dolan, A.M. (2017) Engaging with the world through picture books. In Scoffham, S. ed., *Teaching geography creatively*, London: Routledge, 30–43.
Eisner, E.W. (2002) *The arts and the creation of mind*. New Haven, CT: Yale University Press.
European Commission. (2009) *Arts and cultural education at school in Europe*. EACEA/Eurydice: Brussels.
Gardner, H. (2011) *Frames of mind: The theory of multiple intelligences*. New York: Basic Books.
Gardner, H.E. (2008) *Multiple intelligences: New horizons in theory and practice*. New York: Basic Books.
Gee, J.P. (2012) *Situated language and learning: A critique of traditional schooling*. London: Routledge.
Goldsworthy, A. (2017) *Andy Goldsworthy: Projects*. New York: Abrams Books.
Gozen, G. and Acer, D. (2012) Measuring the architectural design skills of children aged 6–11. *Procedia-Social and Behavioral Sciences*, 46, 2225–2231.

Gray, T. and Birrell, C. (2015) Touched by the Earth: A place-based outdoor learning programme incorporating the arts. *Journal of Adventure Education and Outdoor Learning*, 15(4), 330–349.

Heathcote, D. and Bolton, G. (1994) *Drama for learning: Dorothy Heathcote's mantle of the expert approach to education. Dimensions of drama series*. Portsmouth, NH: Heinemann.

Hicks, D. (2014) *Educating for hope in troubled times: Climate change and the transition to a post-carbon future*. London: Institute of Education Press.

Inwood, H. (2013) Cultivating artistic approaches to environmental learning: Exploring eco-art education in elementary classrooms. *International Electronic Journal of Environmental Education*, 3(2), 129–145.

Kelly, A.J. (2017) Geography and music: A creative harmony. In Scoffham, S. ed., *Teaching geography creatively*, London: Routledge, 163–176.

Mackintosh, M. (2017) Representing places in maps and art. In Scoffham, S. ed., *Teaching geography creatively*, London: Routledge, 76–87.

O'Neill, C. (1995) *Drama worlds: A framework for process drama*. Portsmouth, NH: Heinemann.

Owens, P. (2017) Geography and sustainability education. In Scoffham, S. ed., *Teaching geography creatively*, London: Routledge, 163–176.

Parkinson, A. (2009) Think inside the box: Miniature landscapes. *Teaching Geography*, 34(3), 120–121.

Pickering, S. ed., (2017) *Teaching outdoors creatively*. London: Routledge.

Pike, S. (2017) Creative approaches to learning about the physical world. In Scoffham, S. ed., *Teaching geography creatively*, London: Routledge, 104–118.

Price, L. (2015) Knitting and the city. *Geography Compass*, 9(2), 81–95.

Rauhala, O. (2003) *Nature, science and art*. Helsinki: Otava Publishing Company.

Romey, W.D. (1987) The artist as geographer: Richard Long's earth art. *The Professional Geographer*, 39(4), 450–456.

Roosen, L.J., Klöckner, C.A. and Swim, J.K. (2018) Visual art as a way to communicate climate change: A psychological perspective on climate change-related art. *World Art*, 8(1), 85–110.

Scoffham, S. ed., (2017) *Teaching geography creatively*. London: Routledge.

Sousa, D.A. (2006) How the arts develop the young brain: Neuroscience research is revealing the impressive impact of arts instruction on students' cognitive, social, and emotional development. *School Administrator*, 63(11), 26–32.

Tanner, J. (2017) Geography and the creative arts. In Scoffham, S. ed., *Teaching geography creatively*, London: Routledge, 147–162.

Van Boeckel, J. (2009) Arts-based environmental education and the ecological crisis: Between opening the senses and coping with psychic numbing. In Drillsma-Milgrom, B. and Kirstina, L. eds., *Metamorphoses in children's literature and culture*, Turku, Finland: Enostone, 145–164.

Whitburn, N. (2017) Landscapes and sweet geography. In Scoffham, S. ed., *Teaching geography creatively*, London: Routledge, 88–103.

Whittle, J. (2017) Geography and mathematics: A creative approach. In Scoffham, S. ed., *Teaching geography creatively*, London: Routledge, 131–146.

Whyte, T. (2017) Fun and games in geography. In Scoffham, S. ed., *Teaching geography creatively*, London: Routledge, 12–29.

Williamson, C. and Hart, A. (2004) *Neighbourhood journeys*. London: Cabe Education.

Witt, S. (2010) Geography and art part 2 happy spaces, happy places – Exploring pupils' personal geographies using scrapbooking Summer 18–19. *Primary Geographer*, 72(2).

Witt, S. (2017) Playful approaches to learning out of doors. In Scoffham, S. ed., *Teaching geography creatively*, London: Routledge, 44–57.

8 Powerful geography

Teaching citizenship, global learning and the Sustainable Development Goals

Introduction

The emergence of a postmodern society, with its many unresolved economic, technological, moral and ecological problems, is a global phenomenon. There is widespread consensus that the model of development promoted through globalisation and capitalism is unsustainable (Huckle, 2010). Simply put, the ecological systems and resources upon which we rely cannot sustain current levels of global consumption. Faced with challenges of climate change, rising costs of energy and an increasing population on target to reach 9.5 billion by 2050, the planetary warming and resulting climate disruption will reach unmanageable levels.

Systems of industrial and technological development will have to be reimagined and recreated in ways that do not rely on non-renewable resources, coming to use natural resources in sustainable ways that do not cause harm to people or the natural world. The need to understand the challenges facing the Earth has never been greater. Equally, the need for us to take individual and collective action to conserve resources, to educate others and reduce the impact of destructive forces is paramount.

Growing up in a globalised world requires new approaches to education which develop a global dimension, a futures perspective, and 21st-century competencies. Global learning involves developing an understanding of contemporary issues, local/global links and power relations. By fostering optimism, personal agency and opportunities for action, global learning is empowering, dynamic and challenging. It is facilitated through the development of 21st-century competencies which include critical thinking, communication, collaboration, creativity and innovation.

Within the dynamic between globalisation and education, important debates are occurring and new discourses are emerging. The Sustainable Development Goals (SDGs) formulated by the United Nations include a goal centred on learners gaining the necessary knowledge, skills and attitudes to promote sustainable development (UNESCO, 2015). These SDGs aim to eradicate global poverty, fight inequalities and tackle climate change through 17 integrated goals and 169 targets covering social, economic and ecological issues facing the world today. With the launch of the Sustainable Development Goals, the debate about global learning has moved in from the margins. Recognising the importance of global competencies in the 21st century, the Organisation for Economic Co-operation and Development (OECD) is planning to add to the regular assessment of student knowledge and skills, the assessment of global competency (OECD, 2018).

This chapter sets out to:

- examine innovative approaches to global learning, and opportunities presented by SDGs;
- discuss the implications of living in a globalised society and the importance of global competencies as a set of requirements for living in and participating in a dynamic rapidly changing world;
- illustrate how some schools are approaching global learning in an innovative and inspiring manner.

Globalisation and global competency: Living in a global world

Processes of globalisation are felt by everyone albeit in different and unequal ways. Children in particular are profoundly affected by the processes and impacts of globalisation. Some of these influences can be seen through languages spoken in school, the kind of curriculum being taught, choices made by the teacher, the ethnic background of children and staff, and access to technology. Considering that 25% of the world's population is under the age of 14, the impact of globalisation on school children, their teachers and education needs to be recognised.

Children are living in an increasingly interconnected and globalised society. This brings many opportunities and challenges. Processes of globalisation occur across four spheres: the Economy, Politics, the Environment and the Cultural (Peterson and Warwick, 2015).

Environmental globalisation refers to the global nature of environmental changes and problems including climate change, air quality, water conservation and natural resource depletion. Globalisation has led to an increase in consumption which in turn has placed unmeasurable pressures on the environment.

Economic globalisation refers to the interconnectedness of global finance and trade and the increasing prevalence of multinational corporations such as Google, Facebook, McDonald's, Tesco and Starbucks. Some of these corporations have operating budgets larger than the Gross Domestic Product (GDP) of some countries. In terms of operating budgets, eBay is bigger than Madagascar, McDonald's is bigger than Latvia and Amazon is bigger than Kenya (Trivett, 2011).

Political globalisation The world is more globally connected and therefore decisions are now made in a globally connected setting. Political activity has transcended nation states. Global agencies, international organisations and worldwide political movements have become much more common. Political globalisation refers to the international global system including national governments and intergovernmental organisations. Examples of political globalisation include the European Union, where multiple nations make decisions collectively. Intergovernmental agencies include the International Monetary Fund, World Trade Organization and United Nations. The increase of right-wing politics, the election of President Trump, and Brexit are collectively seen as a backlash against globalisation. Movements against neo-liberal globalisation on the right and left focus on issues such as immigration, austerity, free trade, standards of living and democracy. Responses include intensified protectionism, nationalism, racism and xenophobia or new ways of working together which deal with issues of inequality, poverty and sustainability in the interests of all citizens (Huckle, 2017). New challenges of global warming, social inequality and terrorism are further indicators of how globalised politics may be integral to the shaping of future international policies.

Cultural globalisation refers to the international sharing and spread of cultural norms, features and traditions. Improved transport and technology continue to facilitate cultural globalisation today as people travel and make connections with people in other places. Cultural globalisation has ebbed and flowed during different historical periods. Christian missionaries from Europe brought religious customs and beliefs around the globe. The Silk Road was a trade route between China and the Mediterranean Sea which facilitated the exchange of goods, culture and trade.

Colonisation was a major instigator of cultural globalisation. Commentators such as Spivak (1990) argue that globalisation itself is an extension of the colonisation process. 'Discoveries' of new lands by Christopher Columbus, Vasco da Gama and others expanded trade and economic connections. Transportation of sugar, tobacco, spices, tea, silk, porcelain, gems and precious metals was facilitated at this time. While the historic framing of this story has been in terms of exploration, bravery and heroics, the perspective of indigenous peoples in terms of exploitation, power relations and cultural domination has traditionally not received the same attention. In response to this some American cities have replaced Columbus Day on the second Monday in October with Indigenous Peoples' Day, to move the focus from the conqueror to the conquered. Spivak (1990) describes the process of normalising the myth of Western supremacy as 'worlding of the West as world'. The result of this process is that the contribution of colonisation to an unequal relationship between coloniser and colonised, the resulting exploitation of the Global South and the creation of wealth in the Global North, are completely ignored. The myth of Western supremacy has become the dominant narrative, one which needs to be challenged and unlearnt according to Spivak.

While globalisation is associated with innovation and higher living standards, it also contributes to economic inequality, discrimination and social division. Several well-known brand names have been linked to child labour, unfair labour conditions and unjust practices. Many companies continue to fund cheap labour and sweatshops in the production of their goods. Child labour remains one of the biggest issues of our time. Hence, the importance of exploring questions dealing with the origins and journey of our daily purchases. Developing skills to recognise and question the media and marketing landscape are required. Levels of inequality continue to grow. According to an Oxfam report (2017) 8 men own the same wealth as the 3.6 billion people who make up the poorest half of humanity. With the spread of economic and cultural globalisation, the commercialisation of childhood has become a worldwide phenomenon. Living in an environment saturated by advertising, children are increasingly brand conscious. Sport in particular is a highly brand-conscious sector, targeted by expensive brands such as Adidas, Canterbury, Nike or Reebok.

Global competence

In light of our globalised society, education systems should aim to prepare young people to live and work in a world that is increasingly interdependent. Learning to participate in an interconnected, complex and diverse world (locally and globally) is no longer a luxury but an absolute necessity. The centrality of global competence is well established. Notably, the OECD/PISA (2018: 4) define global competence as

> the capacity to examine local, global and intercultural issues, to understand and appreciate different perspectives and world views of others, to engage in open appropriate and effective interactions with people from different cultures, and to act for collective wellbeing and sustainable development.

Global competence involves possession of the knowledge, skills and dispositions to understand and act creatively on issues of global significance (Mansilla and Jackson, 2011). For instance, examining a global issue requires disciplinary knowledge, the skills to transform this knowledge into deeper understanding and the values and attitudes to reflect on this issue from multiple cultural and diverse perspectives. Globally competent young people investigate the world, recognise a variety of perspectives, communicate ideas and take action.

Recognising the unique role that schools play in preparing our children and young people to participate in our world, the OECD has developed a framework containing four dimensions to explain, foster and assess young people's global competence. The four dimensions, illustrated in Figure 8.1, are as follows:

Examine local, global and intercultural issues.
Understand and appreciate the perspective and world views of others.
Engage in open, appropriate and effective interactions across cultures.
Take action for collective wellbeing and sustainable development.

Global competency requires skills such as reasoning with information, intercultural and communication skills, perspective taking, conflict resolution skills and adaptability. Attitudes of openness, respect for people from different backgrounds and global mindedness have been identified as key dispositions by the OECD.

Education for global competence

Recognising the importance of global competencies and challenges posed by globalisation, education systems have responded through various adjectival educations under the broad banner of global learning. These include development education, education for global

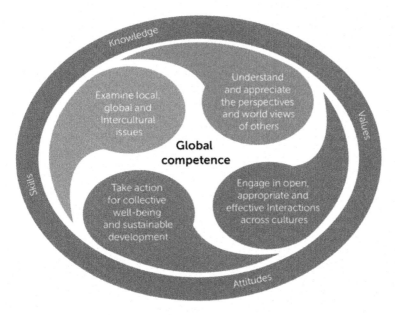

Figure 8.1 The dimensions of global competence
Source: OECD (2018): 11

citizenship, human rights education, and education for sustainable development. These educations value participative, democratic approaches and child agency. Interconnected themes include citizenship, rights and responsibilities, diversity, sustainability, inequality and culture. Key themes and competencies are listed in Tables 8.1 and 8.2. Education for global competency and related educations involve teaching and learning about the wider world and our place in it. According to the OECD (2018) 'these models share a common goal to promote students' understanding of the world and empower them to express their views and participate in society'. While globalisation, global learning and education for global competency are intrinsically linked with geographical education, their importance permeates the whole curriculum. There are several global learning initiatives in schools. In Great Britain, for example, the Global Learning Project (GLP) is a government-funded programme of support helping teachers in primary, secondary and special schools to deliver effective teaching and learning about development and global issues at Key Stages 2 and 3. Local networks provide training, funded continuing professional development (CPD) and curriculum resources as part of this initiative.

An increasingly globalised world raises questions about global learning, global citizenship and what constitutes meaningful citizenship. Global citizenship refers to a sense of belonging to a broader community and common humanity. It emphasises political, economic, social and cultural interdependency and interconnectedness between the local, national and global. Citizenship education is firmly based on human rights and its associated values. According to Oxfam (2015: 5) a global citizen is someone who:

- is aware of the wider world and has a sense of their own role as a world citizen;
- respects and values diversity;
- has an understanding of how the world works;
- is passionately committed to social justice;
- participates in the community at a range of levels, from the local to the global;
- works with others to make the world a more equitable and sustainable place.

Global citizenship education aims to be transformative, building the knowledge, skills, values and attitudes (Table 8.1) that learners need to be able to contribute to a more equitable, sustainable, inclusive, just and peaceful world. Geography, as the curricular area which deals most prominently with global and development issues, has an important role to play in helping schools achieve their aspirations for global citizenship education.

Global citizenship issues by their nature are controversial. To be globally competent one must engage with controversial issues (OECD, 2018). Learning how to engage in dialogue with people who have different viewpoints and perspectives is central to the process of strengthening democracy and fostering a culture of human rights. Often considered too difficult to teach, controversial issues are those which generate strong feelings and divide opinions in communities and society at large. They vary from local concerns to global affairs. The proposal to introduce water charges in Ireland created heated debates in many Irish classrooms while the proposal to build a wall between Mexico and the United States is extremely contentious. While these issues do raise pedagogical issues, such as preventing conflict in the classroom, dealing with biased and stereotypical viewpoints and the role of the teacher's value system, children nevertheless have very strong opinions. Creating a safe space to voice these opinions

Table 8.1 Education for global citizenship – key elements as defined by Oxfam

Knowledge and Understanding	Skills	Values and attitudes
Social justice and equity	Critical and creative thinking	Sense of identity and self-esteem
Identity and diversity	Empathy	Commitment to social justice and equality
Globalisation and interdependence	Self-awareness and reflection	Respect for people and human rights
Sustainable development	Communication	Value diversity
Peace and conflict	Cooperation and conflict resolution	Concern for the environment and commitment to sustainable development
Human rights	Ability to manage complexity and uncertainty	Commitment to participation and inclusion
Power and governance	Informed and reflective action	Belief that people can bring about change

Source: Oxfam (2015: 8)

teaches children how to listen, articulate their own opinions and appreciate that not everyone sees the world as they do. It is important for children to learn how to handle questions of value, how to make well-informed judgements, and how to be flexible enough to change as new perspectives evolve. Global citizenship education involves helping children to form and articulate their own considered opinions in a respectful, safe climate (Table 8.2).

Table 8.2 Oxfam Education for Global Citizenship: A Guide for Schools

Global citizenship involves	It is not
Asking questions and critical thinking	Telling people what to think and do
Exploring local global connections and our views, values and assumptions	Only about faraway places and peoples
Exploring the complexity of global issues and engaging with multiple perspectives	Providing simple solutions to complex issues
Exploring issues of social justice locally and globally	Focused on charitable fundraising
Applying learning to real-world issues and contexts	Abstract learning devoid of real-life application and outcomes
Opportunities for learners to take informed, reflective action and have their voices heard	Tokenistic inclusion of leaners in decision-making
All ages	Too difficult for young children to understand
All areas of the curriculum	An extra subject
Enrichment of everyday teaching and learning	Just a focus for a particular day or week
Whole-school environment	Limited to the classroom

Source: Oxfam (2015: 7)

Moving from a charity-based to a justice-based approach

Education holds a key role in promoting a just, equal and sustainable world for current and future generations. Many schools and educators participate in charitable and fundraising events. However, while social justice is a central aspiration of global learning, for many schools a charity mentality seems to be the norm (Hunt and Cara, 2015). The challenge for global education is to move from charitable approaches to justice and solidarity-based endeavours.

Charity is assistance given to those deemed in need; justice is fairness and an equitable distribution of wealth, resources and power; while solidarity represents unity or agreement of feeling or action, especially among individuals with a common interest. Charity-based actions are short term, targeting symptoms rather than root causes, thus promoting band-aid solutions to complex systemic problems. They can distort people's perceptions of other countries or peoples, particularly of those in the 'South' (Simpson, 2016). While charitable endeavours such as fundraising are well intentioned, it is important for teachers to realise that they do not necessarily promote transformative change and may instead reinforce attitudes of 'us' and 'them' and feelings of superiority. Otherwise referred to as the 'Live Aid Legacy', charitable approaches underpin a relationship between 'Powerful Giver' and 'Grateful Receiver' (Darnton and Kirk, 2011: 6). While charity can be the first step towards social justice, it is ineffective when it is used as a short-term, box-ticking exercise.

In the citizenship literature, Westheimer and Kahne (2004) suggest that educational initiatives, curricula and classroom practices are shaped by ideological and political perceptions of citizenship. They identify three different visions of the 'good citizen': the personally responsible citizen; the participatory citizen; and the justice-oriented citizen. Programmes based on the personally responsible citizen are individualistic, whereby actions involve charitable contributions, good deeds and recycling. Initiatives intended to develop the participatory citizen involve participation in organised efforts to assist those in need. The justice-oriented citizen model is the model least commonly adopted according to Westheimer and Kahne (2004). With its explicit focus on injustice, there is a concerted effort to critically assess social, political and economic structures, to challenge injustice, and to address the root causes of global, economic and social problems.

Education for global competence and related educations exist on a continuum from softer approaches such as teacher-led initiatives including fundraising to more critical approaches involving critical thinking, dialogue and reflection. Andreotti (2006) contends that a more critical approach to global citizenship teaching is needed rather than a 'soft approach' which does not challenge people's assumptions, values and actions and how these have been coloured by personal and societal belief systems. Table 8.3 illustrates the links between 'soft' global citizenship (SGC) and a charity mentality, and critical global citizenship (CGC) and a social justice mentality. It highlights the differences of those two approaches in terms of approach and impact.

In order to achieve deep learning one must first learn how to unlearn (Spivak, 2004) through critical and reflective learning. In this case teachers need to unlearn the charity mentality through deconstructing assumptions about 'us' and 'them', about unequal distribution of resources and about systemic processes which maintain unfair power relations. If we accept that knowledge is socially constructed, as discussed in Chapter 2, knowledge can be reconstructed through a social justice lens.

Table 8.3 Soft global learning and critical global learning

	Soft global learning/charity mindset	*Critical global learning/social justice mindset*
Approach (What?)	One-off campaigns, assemblies, theme days, food tasting Charity or fundraising linked to local and global events/needs Moral/emotive focus (caring value) Focus on poverty (reduction of), helplessness or lack of rights	Global learning approaches within lessons/topics as well as one-off events/days Consider and explore local and global issues Knowledge and understanding focus (educational value) Focus on inequality, social justice and rights
Reason (Why?)	'Impulse to help', moral, being 'good' Responsibility FOR the other (or to teach them) Caring for poor people	'Impulse to understand', equity and ethical Responsibility towards the other (to learn/decide with the other) Solidarity with people without rights or opportunities, and challenging this where possible
Action (How?)	Help people to survive poverty – raising money for poor countries overseas Sharing our wealth	Participate in structural change for elimination of poverty and inequality Critiquing how we became wealthy
Learning (Message)	Reduce poverty through charitable work, campaigning and fundraising	Challenge inequality and injustice, and support rights for all
Outcomes (Positive)	Feel-good factor Greater awareness of some of the problems Motivation to help/do something	Sustained engagement Independent/critical thinking, and more informed, responsible and ethical action Encourages pupil voice and advocacy Encourages self-reliance and self-determination for poor countries
Outcomes (Negative)	Can encourage or sustain a sense of cultural superiority or privilege Sustains dependency for poor countries Reinforces prejudice and stereotypes Uncritical action	Sometimes uncomfortable, and can bring about guilt and shame Can feel overwhelming leading to a feeling of helplessness
Ultimate goal (Of education?)	Empower individuals to act (or become active citizens) according to what has been defined for them as a good life or ideal world	Empower individuals to reflect critically on their understandings and perceptions, to imagine different futures, and to take responsibility
Pupil participation (Where does the change happen?)	From the outside to the inside (imposed change)	From the inside to the outside (negotiated change)

Source: Simpson and Barker (2016)

Critical global learning

Critical global learning is urgently needed to give children and school communities hope in troubled times (Huckle, 2017). A critical approach to global learning involves a continual questioning of ideas and assumptions. Critical geographical thinking is discussed in Chapter 2. While critical thinking is an essential 21st-century competency, it is also important in the context of powerful primary geography and global learning. Critical thinking involves active engagement with ideas, it entails looking at issues from different perspectives, it necessitates having reasons for one's opinion and it means thinking independently for oneself (Roche, 2014).

Narratives, stories, news stories and individual commentaries are shaped by the ideologies of those who compose them and by those of society. Van Dijk (2011: 286) describes ideology 'as an interpretative frame through which social actors make sense of their relationship to the world'. However, this ideological framework is not neutral as it colludes in the creation and maintenance of power relationships. Teachers, curricula and schools can sustain or challenge dominant ideological views and power relations. Schools should not be afraid to provide learning environments that involve analysis, action and an understanding of knowledge. Critical pedagogical approaches drawing on the work of Freire, Giroux and Spivak (2004) suggest that both teachers and children need to 'unlearn'. This involves questioning our taken-for-granted assumptions.

Children's stories provide an opportunity for children to consider scenarios, explore alternative endings, imagine future iterations and to think critically. There are numerous picture books available which present geographical themes to children in an age-appropriate way. In teaching children how to interact with the illustrations, the text and the combined message through making their own meanings, children can learn the skill of critical thinking in an age-appropriate manner (Roche, 2014; Dolan, 2014).

Stories provide fertile ground for analysing power relations, class issues and structural inequality. Whose story is being included and excluded can reflect society's dominant ideology. Equally stories themselves can be transformative, using narrative to challenge dominant paradigms. Stories can bring empathy and inspiration to challenging real-life issues. Furthermore, they can reframe the notion of what is possible and impossible.

In 2009 the Nigerian writer Chimamanda Ngozi Adichie gave a noteworthy TED talk called *The Danger of the Single Story* (Adichie, 2009). Adichie is a novelist from Nigeria, who was influenced from a young age by the novels of the Western world. All the books she read were foreign, and she was convinced that books had to be about things with which she could not identify. Then she learnt of African books, books written by Chinua Achebe and Camara Laye. This is when Adichie began to write about things she recognised. When Adichie came to the United States to go to college, her roommate was surprised that she spoke English and liked to listen to Mariah Carey. The roommate expected that Adichie wouldn't know how to speak English, or understand how to use a stove, and she expected to hear tribal music. Adichie's roommate held a single story of Africa. According to Adichie (2009) 'in this single story there was no possibility of Africans being similar to her in any way, no possibility of feelings more complex than pity, no possibility of a connection as human equals.' It was about what happens when complex human beings and situations are reduced to a single narrative: when Nigerians, for example, are treated solely as poor, desperate and pitiful. Her point is that each individual person, each place, represents multiple stories. Reducing a person to one story is an act of dehumanisation. Adichie, in her Ted Talk, (2009), describes the powerful impact of stories:

> There is a word, an Igbo, that I think about whenever I think about the power structures of the world, and it is 'nkali'. It is a noun that loosely translates 'to be greater than another'. Like our economic and political worlds, stories too are defined by the principle of nkali: How they are told, who tells them, when they are told, how many stories are told, are really dependent on power.

During the TED talk she states 'the single story creates stereotypes, and the problem with stereotypes is not that they aren't true, but they are incomplete. They make one story become the only story'. There exists a single story of Africa, but it is only one small part of the story. When people think of Africa, they think of violence, poverty, AIDS, uneducated people, mud huts, helpless people, hopelessness and corrupt governments. Single stories are one-dimensional and often misrepresent complex multi-layered situations. They often become the only stories. Avoiding the single story helps to challenge stereotypical thinking.

Africa is not a country! Yet many people erroneously refer to this diverse continent as one country or one nation. Furthermore, literature for children set in Africa continues to be dominated by stereotypes. While several authors attempt to address the complexities of various cultures, their efforts are often thwarted by publishers whose primary concern is book sales as opposed to authentic representation of cultures. Africa is not a country, Africa is the second-largest continent in the world, three times the size of the United States of America. This vast continent with 55 countries (including the recently established state of Southern Sudan) has at least 800 different cultures and dialects. One-tenth of the world's population live in Africa. Yet the importance of Africa is not reflected in the classroom, the curriculum or in the selection of children's literature which is shared with children. It often surprises children to discover the size of Africa (as illustrated using a Peters Projection world map), to find out that Africa spans seven time zones. It takes longer to travel from Dakar, Senegal to Nairobi, Kenya than to fly from New York to London. There are no tigers in Africa and most African children have never seen some of the wildlife in Africa, other than in a zoo. However, stereotypical images of Africa pervade in terms of wild animals and primitive portrayals of cultures. Few people visualise places in Africa in terms of skyscrapers, busy city streets, business suits and success stories. One of the first books to tell the story of European colonisation from an African perspective, *Things Fall Apart*, was written by Nigerian author Achebe in 1958. This marked a turning point for African authors, who began to reframe the narrative of Africa as the so-called 'dark continent'. There are now many books which provide alternative cultural narratives. Teachers have a much broader selection of books and stories. It now remains for teachers to choose stories wisely, sensitively and empathetically.

Picture books such as *Africa Is Not a Country* help challenge some of the dominant stereotypes which are prevalent today. *Africa Is Not a Country* (Knight et al., 2002) is perhaps the first picture book about the African continent to respectfully present the diversity of people living in several countries. The authors and illustrator have created a book that explicitly describes and illustrates the multiple dimensions of the continent. Referring to 25 countries in total, the book begins with Eritrea and features often forgotten children from Cape Verde, Lesotho, Mauritania and Madagascar. Moreover, the book focuses on children's activities and not those of adults. This book tackles the misconception that

Africa is a country. The author–illustrator team narrate the experiences of children from a range of African countries at play, school and in the home. Realistic illustrations are used to explore the cultural, environmental, ethnic and social diversity of countries which make up the African continent. The book begins with morning and ends at night and depicts children having breakfast, going to school, doing housework, shopping and playing. The emphasis is always on modern people. Unique characteristics for each African country are highlighted. For example, Rwandan children are shown making pictures of war, while two Kenyan children are illustrated as they run to school, dreaming of one day becoming professional runners. Teachers can also show children a range of other images from African countries to supplement the central message from this book. The vast and varied African continent is shown using maps and a variety of people inhabiting different environments. From vast deserts with camels in the north, to lush agricultural lands in Central and Southern Africa, Africa is introduced to young children in this colourful and easily readable book, with the explanation that Africa is so large, diverse and complex that it should not be thought of as a single nation. Africa has so much to offer – soccer, agricultural products, and different religious faiths – and the diversity is not only of land and culture, but of people too. Each country is presented in a two-page spread, with text and a large illustration. The text includes many facts about each country, without being overly detailed. At the end of the book is an alphabetical text box for all African countries (excluding the recently established state of Southern Sudan), featuring capital city, population, Independence Day, currency, a pronunciation guide, national flags, and unique facts about the country. Beautifully illustrated and well researched, it will be a joy to young children being introduced to Africa and the many countries that make up the African continent.

Countering stereotypes: Learning how to unlearn

A *stereotype* is a belief about or a perception of an individual based on the real or imagined characteristics of a group to which that individual belongs. Stereotypes can lead us to judge an individual or group positively or negatively. Even stereotypes that appear to portray a group positively reduce individuals to categories and tell an incomplete or inaccurate 'single story' (Griffiths and Allbut, 2011). *Prejudice* occurs when an opinion about an individual or a group is formed based on a negative stereotype. When a prejudice leads us to treat an individual or group negatively, *discrimination* occurs. Stereotypes shape single stories which can become deeply embedded in our world view and attitudes to other people and places. Once formed these single stories can limit our thinking, reduce our capacity for criticality and perpetuate the original stereotype.

Children form their ideas about other places from multiple sources. They are immersed in a social, cultural and political environment which produces and perpetuates stereotypical images. Social media, TV shows, books, advertisements, friends and extended families all communicate images, values and perspectives explicitly and implicitly. These messages influence the way that children develop and engage with the world. Research by Oberman et al. (2014) illustrates stereotypical views of 9-year-old Irish children which clearly have been informed by charity appeals. This research also suggests that introducing a global perspective into early childhood education, using open-ended and active methodologies, supports the development of global citizenship skills, attitudes and understanding. A central aspect of primary geography, global education provides a broader, more nuanced view of other people and places. It also provides children with the skills to recognise and challenge stereotypical images.

Questions to discuss are:

What stereotypes can we identify in our view of Kenya (or Africa) and its people?
Why are these stereotypes so persistent?
How accurate are they?
How do we know?
How might they affect the lives of people living in the West?
Why do we stereotype other cultures?
How is stereotyping linked to prejudice about others?
Can we identify stereotypical views held by others about our own country?
Are they accurate?
Are they positive or negative?
Where do they come from?

(Dolan, 2014: 57)

Education for sustainability

While we have learnt to live unsustainably in a very short space of time (Inman and Rogers, 2006) the challenge for educators is how to promote learning for sustainability in the interests of the future wellbeing of the planet and ourselves. To live sustainably demands nothing less than a complete change in our relationship with others and the world, a change in our way of being, something we might call an 'ontological epiphany' (Wade, 2006). This is challenging and in some cases disturbing as the very structures through which we live, work and commute encourage us to live unsustainably. A fundamental change in our core values and attitudes is required. Kumar (2005) and Selby (2007) suggest that the psychological or spiritual dimensions of change are just as important or perhaps more important than the political dimension; they are informed by core values which ultimately form the basis for our actions. Several commentators argue that modern society supports unsustainable development through values and ideas communicated through mass media and economic, social and political institutions (Bonnett, 2008; Huckle, 2017). Sterling (2012) argues that our education paradigms need to be re-oriented and refocused towards sustainability as a matter of urgency. The Irish Government has published its first National Strategy on Education for Sustainable Development (DES, 2014). The strategy, entitled Education for Sustainability (EfS), aims to

> ensure that education contributes to sustainable development by equipping learners with the relevant knowledge (the 'what'), the key dispositions and skills (the 'how'), and the values (the 'why') that will motivate and empower them throughout their lives to become informed citizens who take action for a more sustainable future.
>
> (DES, 2014: 3)

Interestingly this strategy was not afforded the same priority, commitment and funding as that given to the literacy and numeracy strategy. To date EfS has not been embedded as a key priority within education programmes, nor has it been mainstreamed into teacher education (Dolan, 2012). Notwithstanding the lack of political commitment for EfS, the Sustainable Development Goals (SDGs) (discussed later in this chapter) provide a framework for threading citizenship, sustainability, human rights and global learning throughout the entire curriculum.

Case study 8.1 What does sustainability look like? Visit to Cloughjordan ecovillage

The realisation that climate change generates environmental havoc around the world is bringing more attention to communities focused on environmental resilience and sustainability. It is important for primary teachers wishing to teach sustainability issues to experience models of living sustainably during their pre-service or in-service education. One such example of sustainable living is the Cloughjordan ecovillage situated in north Tipperary. Every year my geography specialism students spent some time in the ecovillage to explore options for their probable and preferred futures (Hicks, 2009) (Figures 8.2 and 8.2a).

Approximately a third of the 67-acre site is residential property, a third incorporates community woods and a third is dedicated to the community farm. On the site, there are 55 low-energy homes, work units, an enterprise centre, a bakery, a pottery studio, a café and a hostel. There is also a wood-powered community heating system, a green enterprise centre, an eco-hostel for tourists and access to a railway station. The community farm produces food for the villagers and broader community. Sixteen acres of broadleaved forestry and a meandering biodiversity trail, locally known as the 'perimeter walk', promote biodiversity. The village is well integrated into its surrounding area and has helped regenerate part of rural Tipperary.

There is a very strong ethos of education whereby the village, as a model of sustainability, shares its practice near and wide. There is a steady trail of visitors and tourists including children, who come to learn in the community. As well as achieving a low carbon footprint, the village also offers courses in skills like permaculture, design principles, eco-building and horticulture.

Working cooperatively is a key principle and the community farm is an interesting example. Members pay an annual fee in order to purchase products from the farm, as well as contributing personal time when they can. A sense of community and reliance is at the heart of this project. At the national level the community's ecological footprint is the lowest measured in Ireland. The ecovillage challenges the student teachers' perceptions of sustainable living and encourages them to reflect on their own personal and professional practice. The experience has a deep impact on the students, as illustrated in their reflections:

> The visit to the Eco-Village can show student teachers that sustainable livelihoods are attainable and do not require massive life altering sacrifices that will leave you deprived in society. Cloughjordan has an abundance of qualities that allow teachers to create innovative classroom lessons and provide learning opportunities for students of all ages. Most importantly, the visit showcased the links between community, companionship and wellbeing.

> My visit to Cloughjordan has taught me that teaching about sustainability and climate change cannot be tackled through the school textbook alone. Our students must experience sustainable behaviours for themselves through hands-on learning. It is vital that students are taught the skills necessary to deal with future dilemmas their generation will face at the hands of climate change.

> As a teacher inspired by the cooperative approach applied in Cloughjordan, I hope to promote more co-operative movements within the schools I teach in the future and

Figure 8.2 Student teachers' visit to an ecovillage

Figure 8.2a Student reflection on visit to the ecovillage

spread the message of their initiative. I will instil the message in my students to the value of community and how working towards a common goal can positively affect the wellbeing of every person and the environment they live in.

Much of the discourse and research literature about climate change is rooted in a pessimistic space; very often instilling a sense of hope for the future is neglected. If we are to teach our children about climate change effectively, it is essential that we provide them with practical solutions to fight against climate change while also ensuring that they remain hopeful and optimistic about the future. Cloughjordan eco village provides the student teacher with an insight into a community determined to fight against climate change, a community who support one another and the local economy and a community who instil a sense of hope for the future in those that visit them, such hope can be then transferred to our children in classrooms.

My biggest takeaway from the visit was witnessing the ethos of the village in terms of their community spirit and their outlook on the power of sustainable living and its impact on not just climate change but our own wellbeing and community spirit.

It is without doubt an experience like none other for to understand the message encapsulated within the village one must walk the woodland paths and speak to the locals and witness the community ethos by walking through their farm, eating lunch in their cafe and walking through the food stalls in which they collect their fresh produce.

Prior to ever visiting such a place one may begin to think of a radical society that is completely new and different to anything we've ever experienced. Whereas upon visiting you begin to discover that this way of living and the societal connection between the producers and consumers is based upon the values of a sustainable society as these business owners care about their locality rather than profiting from their product through market competition and expansion.

The Green Schools programme

Green Schools is an international environmental education initiative offering a structured programme and award scheme for schools to take environmental issues from the curriculum and apply them to the day-to-day running of their school. The Green Schools programme is run by FEE (Foundation for Environmental Education), and is known internationally as 'Eco Schools'. Green Schools are now represented in over 53 countries all over the world.

Offering a holistic approach for schools, the programme aims to make education for sustainability part of the life and ethos of schools and by extension the wider community. The Green Schools award takes the form of a Green Flag, now established as a well-respected and recognised eco-label.

'Eco Schools' is an initiative of FEE (Foundation for Environmental Education), and is referred to by a number of different names in member countries, e.g. *Eco-Scholen* in Belgium, Eco Schools in the UK, Green Schools in Ireland and *Umweltschule* in Germany (see www.eco-schools.org for more information). 'Eco Schools' began in 1994 as a pilot project in the UK, Denmark, Germany, France and Portugal. In 2001, 'Eco Schools' transformed from a European initiative to one of international dimensions with the inclusion of Africa, South America, Oceania and Asia.

The main aim of the programme is to motivate school-aged children to become pioneers of change, and to find and implement solutions to environmental and sustainable development challenges which are encountered at a local level. Children are incredibly resourceful and the Green Schools programme provides a productive outlet to channel that resourcefulness in their own communities. The main idea behind the Green Schools programme is that green or eco-thinking should be mainstreamed. In this way future decision-makers, politicians, consumers and citizens will be more sensitive to the fragile nature of our environment and will take action to ensure its wellbeing.

Following the appointment of a Green Schools committee, each school begins its Green School programme with a review or assessment of the environmental impact of the school's activities. The results of the school's Environmental Review help the committee agree an action plan. This helps the school community decide whether change is necessary, urgent or not required. Schools work on a number of themes: waste management, waste minimisation, school soils, biodiversity, energy, water, transport, health and wellbeing, and sustaining the world. The Green Schools Expo is a national exhibition which showcases some of the most innovative projects from primary and secondary schools in Ireland. Children proudly present their projects and their passion and enthusiasm is

Figure 8.3 Green Schools Expo: national exhibition of the Green Schools Project

evident. Figure 8.3 and Case Study 8.2 illustrate some of this work. These projects illustrate many of the features of powerful primary geography discussed in this book including critical thinking, creativity, collaboration, team work and real life problem-solving.

Case study 8.2 Green Schools initiative: Esker Educate Together National School, Lucan, Co. Dublin

This is a 16-class school with 27 teachers. The school has already been awarded a Green Flag based on the themes of litter, waste and energy. The school is currently working to achieve its third Green Flag, based on the theme of water. The children are very enthusiastic about their work and this passion is the driving force behind their work. The Green Schools committee is made up of children from each class from juniors to seniors. Committee members passionate about the programme are selected by their peers. Each member is required to make a presentation to their class about how they can contribute to the programme prior to being elected. This year the committee members are called 'Water Warriors', as they are focusing particularly on reducing water usage in the school and stopping water pollution in their area. Following a review of the school's initial situation, the committee took action to reduce water consumption in their school. They have a tank on their roof which collects rain water that is recycled in the toilets around the school. The committee came up with three simple ideas to reduce water wastage in their classrooms:

1 Put the plug in the sink when washing paintbrushes.
2 Small flush for one, big flush for two.
3 One push of the tap is enough to wash your hands.

The school also organised a 'Walk for Water', which aims to spread awareness of water issues around the school. The children walked 5km, carrying bottles of water on their back, to help them understand that some people carry water for survival. Although this year the school's focus is on water, the children continue to carry out other actions from previous themes such as litter picking and turning off the lights.

Unfortunately, while monitoring the school's water usage, the committee discovered their water usage had increased in previous months. The children were deeply disappointed

Figure 8.4 The Community/School Garden: Esker Educate Together NS Lucan, Co. Dublin

as they had worked so diligently to reinforce their message across the school community. However, they found a water leak. Due to their diligence the pipe was repaired, and the school was saved from unnecessary water wastage. The school incorporates the Green Schools programme into various curricular areas, with a strong focus on the arts, as discussed previously in Chapter 7.

The school also hosts a wonderful gardening club recently crowned 'Winner of the Community/School Garden or Allotment Award'. The gardening club has been instrumental in changing what was a bland piece of land to the beautiful, colourful, interesting and enticing space that it is now. Their magical fairy village is an ongoing project. The intercultural aspect of this project is most inspiring as it creates a space for parents from different ethnic groups to work together on a common cause, building relationships, practising their English and providing an enhanced environment for their children (Figure 8.4).

Case study 8.3 A focus on plastic

The Collins dictionary 'Word of the Year' 2018 award was bestowed on the term 'single-use'. Single-use plastics, also referred to as disposable plastics, are commonly used for packaging items. These include, plastic drinking bottles, grocery bags, food packaging, straws, containers, cups, lids, cutlery and foam takeaway containers. While single-use may have once epitomised a carefree and convenient lifestyle, it has now come to represent society's worst excesses. Since the revolutionary invention of plastic in the 1950s, shopping and consumption have changed dramatically. Plastic has pervaded our world because it is cheap, durable and lightweight. Programmes such as the BBC's *Blue Planet II* have heightened public awareness about the damage caused by plastic packaging. Images of plastic adrift in the ocean, causing the suffering and death of marine animals, have led to a global campaign to reduce its use. Nevertheless producers and consumers still opt for low-cost plastic products. Mysteries such as those cited in chapter 2 (Table 2.8: Mystery: Where did Orla's Barbie end up?) provide an opportunity for children to think about the journey of plastic and the power of the consumer.

Several schools are researching the use of plastic in their school and locality. St Clare's NS, Harold's Cross, Dublin has 330 children and 22 teachers. A previous winner of the Dublin City Neighbourhood Award, the school holds seven Green Flags. These were obtained by reducing energy and waste in the school, improving the school environment, planting new trees and creating a butterfly garden. St Clare's NS exemplifies a highly efficient Green School that runs ongoing initiatives and competitions to support an eco-friendly lifestyle. The committee consists of 8 children between the ages of 9 and 13, selected randomly. The committee meets once every fortnight to discuss projects and share ideas. The committee presented its project *#the plastic challenge* at the Green Schools Expo in the Royal Dublin Society (RDS) exhibition space in Dublin. The Plastic Challenge is a whole-school campaign to boycott the use of plastic bags in local shops. The local SuperValu listened to their consumer voices, recognised the plea in their letters and stopped providing plastic bags for fruit and vegetables. Their success highlights that consumers hold the power to change corporate business and multinational company policies.

The 'Plastic Elephant' was an experiment conducted by 10-year-old children to unveil the amount of plastic the school consumes weekly. Each child was asked to bring in the plastic consumed at home for one week. The children were well informed about Plastic Island, also known as the Great Pacific Garbage Patch, where more than 38 million pieces of unrecyclable material is dumped. This was an issue of considerable concern to the children, particularly the welfare of plant and animal life in the ocean. The plastic brought in by the 330 children equalled the weight of an adult Asian elephant. The project raised awareness of the overconsumption of plastic in our daily lives, highlighting the severity of the 'Plastic Island' epidemic.

A video tutorial on the school website showcases the children making eco-friendly bags out of recycled newspaper. These bags are used in a local flower shop. This activity utilises the children's skills while also raising awareness of the plastic bag crisis. The project fosters creativity, innovation and team work. It is evident that the children are motivated to make a difference locally and globally for a greener, eco-friendly and sustainable world.

Children from St Brendan's NS Blennerville, Tralee Co. Kerry created an art installation based on ocean pollution (Figure 8.5). It included a crocheted coral reef before and after pollution. The following is an extract from the school new letter:

A Plastic Ocean by a nine year old child

The Green School Committee has been concentrating on the bad effects that plastic is having on the oceans and the creatures that live in the oceans. Plastic rubbish appears in every ocean of the world and every year we are adding millions of tons more. Plastic can't break down in the sea water and can cause injury or death to marine life. A lot of plastic is only used once then thrown away. We want to reduce the once-off use of plastic and everyone in school can help by stopping the use of one-use only drink containers such as Capri-Suns, plastic bottles and juice cartons. Straws are another thing we could stop using. The U.S. alone uses 500 million straws each day. They are generally made from single-use plastic and are so flimsy that most of them can't be recycled. So can we all make the effort to stop using one-use only drink containers and think about sea-life before you take a straw.

230 *Powerful geography*

Figure 8.5 Plastic ocean project

A focus on forests: The leaf project

Forests are complex ecosystems, providing wildlife habitats as well as much-needed recreational facilities. Throughout the centuries in Ireland, trees have played vital, practical and spiritual roles in the lives of Irish people. Ireland was once covered by a blanket of beautiful oak, elm and pine forests. However, due to a combination of bog expansion and the arrival of Neolithic farmers 6,000 years ago, the density of our forests declined. Today Ireland has the lowest forest cover of all European countries: approximately 11% compared to an European average of well over 30%.

Timber is one of Ireland's very few renewable natural building materials. It is a sustainable, renewable and a home-grown source of heating and electricity. Trees can contribute to the landscape in many ways: in hedges, small woodlands and forests. They also store large amounts of carbon which is essential in the context of challenges posed by climate change discussed in Chapter 5.

The Learning about Forests (LEAF) programme enables children, teachers and student teachers to explore the functions of forests. The programme includes topics such as biodiversity, water, climate change, timber products and community. LEAF advocates outdoor learning and hands-on experience which help children to acquire a deeper and more engaged understanding of the natural world. While the focus of the LEAF programme is on tree-based ecosystems, the skills and knowledge can be applied to any natural environment or ecosystem. LEAF is an international programme initiated by the Forest in Schools programme in Norway, Sweden and Finland in cooperation with the Foundation for Environmental Education (FEE).

In Ireland, Curragh Chase Forest in County Limerick is used as the South-West site for the national LEAF programme. The park includes 313 hectares of mixed woodlands, park land and lakes, providing a rich habitat for a diverse range of animals and plants. Initial teacher education students have a chance to visit this forest and work with the LEAF facilitators (Figure 8.6). Here is a selection of their reflections on their visit to the forest.

> Our class trip to Curraghchase was enlightening. From our time in the park we learnt a variety of different ways that the outdoors can be used to educate children not only in the subject area of geography but in other curricular areas. Additionally, I learnt that a deep knowledge of nature is not required to teach the children. Both the children

Figure 8.6 Student teachers on a field trip to Curraghchase Forest, Limerick

and teachers can learn together through the interaction with nature. All of the activities that we engaged in on the trip are relevant to school placement. The measuring of tree height and also the age of the tree is extremely relevant as it integrates the subjects of geography and maths and can be used for various class levels. One activity that I particularly enjoyed was the 'Web of Life'. This activity is inclusive and allows the children to express their understanding on opinions. There is no denying that outdoor learning is fundamental in the child's holistic development. The children learn about their surroundings and can engage with the environment they live in.

I learned a great amount from my visit to Curraghchase Forest. Firstly, I learned about the different categories of trees that exist in this forest habitat. These include native trees, non-native trees and invasive trees. I enjoyed listening to the history of this forest and the interesting features on the grounds of Curraghchase, such as the main house that is now home to a large bat habitat. I also enjoyed learning about the LEAF project, and it is something that I would like to research more in the future.

An activity that inspired me during this visit was the activity regarding the making of pendants out of pieces of wood. I think children would really enjoy this and it shows how nature can be linked throughout the curriculum, especially through the arts.

There is so much value in outdoor learning. For example, I learned first-hand how forests create and manage themselves. I saw the different ways that trees compete to survive, and how their competitiveness can influence the way that they grow and how they position themselves in the habitat. I learned that outdoor learning supports peoples' well-being, with the outdoors being a place where children can be allowed to be still, calm and reflective. Also, being in such an impressive woodland environment such as Curraghchase would be a valuable outdoor experience for children who are from more built up city areas.

At one point we all sat down and Ray gave each of us a small piece of timber to hold onto. I immediately thought of a range of teaching opportunities a simple piece of timber could provide. Children learn very well through physical interaction and so to give each child a piece of the forest to hold and keep would not only enhance their learning, but also act as a beautiful reminder of their forest walk. He told us stories about how some children react on these walks as they have never walked in a forest before. Having grown up on a mountain with a vast forest just behind my house, I was truly shocked to hear this. It caused me to reflect on how my experiences of walking and playing in my local forest growing up influenced my relationship with nature. If a child does not have such experiences because of where they live for example, they might struggle to understand the importance of trees and nature as a whole throughout their life. As a teacher I would like to provide children with opportunities to explore nature in a fun and educational way, something I had not considered much before this trip.

Forest school

Forest School is an opportunity for a group of learners and leaders to spend a sustained period outdoors, once a week, in a wooded environment, ideally year-round. A regular routine is followed that is learner-led and facilitated by trained leaders. Forest School involves children having regular contact with woodlands over an extended period of time (Figure 8.7). It allows them to become familiar with and have contact with their natural environment. While the concept is increasingly popular in Great Britain, the Irish Forest School Association has recently been established. The Forest School movement facilitates innovative outdoor learning and play. An inspirational process, Forest School offers all learners regular opportunities to achieve and develop confidence and self-esteem through hands-on learning experiences in a woodland or natural environment with trees (Knight, 2013). Its philosophy is based on developing children's motivation, emotional and social skills, resilience and self-awareness through positive outdoor experiences.

Figure 8.7 Getting to know your forest

Figure 8.8 Forest School: St John the Apostle NS Knocknacarra, Galway

During Forest School, children engage in activities such as building shelters, cooking on campfires and identifying plants and wildlife. They use tools to create artworks, they listen and respond to stories. Working in teams they work collaboratively to complete tasks and solve problems. Children use their own initiative and learn how to handle risks. They learn about the woodlands as a vibrant and diverse habitat. This holistic programme facilitates the development of the whole child by fostering independence and self-esteem through engagement with the natural environment. Learning is holistic and closely related to each child's developmental stage and curricular requirements. Educators wishing to become Forest School Leaders can acquire a qualification, which includes obtaining an outdoor first aid certificate. Trainee Forest School Leaders conduct a six-week block of practice and must submit a portfolio of work that is approved by an accrediting body. Some schools with teachers who have acquired Forest School training adapt aspects of the Forest School programme for their own purposes (Figure 8.8).

The global goals for sustainable development

The Sustainable Development Goals (SDGs) or Global Goals agreed by world leaders are a set of 17 ambitious targets to end poverty, protect the planet and ensure prosperity for all (Figure 8.9 and Table 8.4). The development of the Global Goals involved the biggest consultation process ever undertaken in UN history. Building on the achievements of the Millennium Development Goals, these new goals apply to every country (albeit with some goals being more applicable than others in certain cases) and provide the focus for all development work over the next 15 years. Since their launch in September 2015, the UN's Sustainable Development Goals (SDGs) have been adopted by schools around the world. The SDGs provide a unique framework for teachers for thematic learning, global learning, sustainability and citizenship education.

While the SDGs offer a 'major improvement' over their predecessors, the Millennium Development Goals (MDGs), they have been criticised for promoting traditional models of industrial development and a 'business as usual' response to the challenges facing the world today (Sterling, 2015). The goals are wide-ranging and aspirational, but they also exhibit certain contradictions. There is some concern that the goals are presented in

234 *Powerful geography*

Figure 8.9 The Sustainable Development Goals infographics

Table 8.4 The Sustainable Development Goals

1. End poverty in all its forms everywhere.
2. End hunger, achieve food security and improved nutrition and promote sustainable agriculture.
3. Ensure healthy lives and promote wellbeing for all at all ages.
4. Ensure inclusive and equitable quality education and promote lifelong learning opportunities for all.
5. Achieve gender equality and empower all women and girls.
6. Ensure availability and sustainable management of water and sanitation for all.
7. Ensure access to affordable, reliable, sustainable and modern energy for all.
8. Promote sustained, inclusive and sustainable employment, full and productive employment and decent work for all.
9. Build resilient infrastructure, promote inclusive and sustainable industrialisation, and foster innovation.
10. Reduce inequality within and among countries.
11. Make cities and human settlements inclusive, safe, resilient and sustainable.
12. Ensure sustainable consumption and production patterns.
13. Take urgent action to combat climate change and its impacts.
14. Conserve and sustainably use the oceans, seas and marine resources for sustainable development.
15. Protect, restore and promote sustainable use of terrestrial ecosystems, sustainably manage forests, combat desertification and halt and reverse land degradation and halt biodiversity loss.
16. Promote peaceful and inclusive societies for sustainable development, provide access to justice for all and build effective, accountable and inclusive institutions at all levels.
17. Strengthen the means of implementation and revitalise the global partnership for sustainable development.

'silos', for example, goals addressing challenges such as climate, food security and health are presented in isolation from each other with no cross-references or links to an integrated plan. The lack of integration may generate conflict between different goals, most notably trade-offs between economic growth and a move towards sustainability. While the language is transformative, the goals promote a pro-growth model of development and a utilitarian approach to education (Brissett and Mitter, 2017). The goals, and in particular SG4, are inconsistent with the philosophy of transformative education which challenges the very social and economic fabric of contemporary society. Notwithstanding well-founded criticisms, the goals provide considerable opportunities for education in general and global learning in particular. Education is a mechanism for achieving other goals, such as good health and wellbeing (Goal 3) or climate action (Goal 13). It is also potentially a mechanism for encouraging learners to engage critically with the goals and the understandings of sustainable development they promote. A number of excellent resources are available for teaching the SDGs creatively and critically.

World's Largest Lesson (WLL) is a collaborative education project to encourage teaching of the United Nations Global Goals for Sustainable Development. The World's Largest Lesson aims to facilitate all schools from around the globe to teach their children about these goals. The WLL includes lesson plans, animated films and accessible guides to the goals. Teaching about the SDGs is becoming a global trend on social media and is connecting teachers and schools around the world. While the global challenges are predominantly geographical in nature their interdisciplinary appeal facilitates collaboration, cross-curricular links and creative responses. Teaching about the SDGs provides a useful entry point to teaching sustainable development issues on local, national and international levels. The fourth SDG deals with quality education, lifelong learning and access to education for all. Target 4.7 is as follows:

> By 2030, ensure that all learners acquire the knowledge and skills needed to promote sustainable development and sustainable lifestyles, human rights, gender equality, promotion of a culture of peace and non-violence, global citizenship and appreciation of cultural diversity's contribution to sustainable development.
>
> (UN, 2015)

Case study 8.4 Irish Aid Awards

The Irish Aid Our World Awards is an annual awards programme for Irish children from 9 to 13 years of age. Children work in pairs, as a class or as a whole school to present a project exploring the United Nations Sustainable Development Goals and the work of Irish Aid.

The Our World Awards seek to enable children to learn about the lives of children and their families in the Global South and how Ireland, through Irish Aid, and 192 other countries through the United Nations, are working together to create a safer and fairer world and a better future for all the world's children. Children as global citizens can play their part by taking part in the Awards and engaging with these issues.

Sixty primary schools from across Ireland are selected from hundreds of entries to showcase their projects in four regional finals ahead of the national final in June. The four regional events take place in Sligo, Limerick, Cork and Dublin. Twelve schools from the regional finals go forward to the national finals in Dublin Castle. Projects feature the realities of life for children and their families in Malawi, Ghana, Cambodia, Tanzania and many other countries. Children use a variety of creative methods to communicate what they have learnt including illustrated project books, film, artwork, music, board games and interactive websites (Figure 8.10). They also present their projects to others in their wider school and local communities. Feedback from participating teachers and children has been very positive.

We found that our school's participation in the Irish Aid One World competition to be a very worthwhile and engaging project for all involved. It was an effective means of raising awareness of Global Issues at classroom level.

Two of our 6th class girls took the initiative to do their own additional research. Their enthusiasm, in turn, motivated their peers to engage in finding out more about the subject. Consequently, the class gained a deeper understanding of the issues involved. As a result of one class participating in the project, the overall awareness of Global Issues throughout the school increased.

An added bonus was the fact that the children got to go and present their project in Mary Immaculate College Limerick, where they met children from other schools and saw how others presented projects on the theme.

The children who were involved were proud of the legacy they left with the school, as having participated in the project we have a plaque displayed in the school acknowledging their participation.

Principal, St John the Apostle NS, Knocknacarra

The Irish Aid World Awards was hugely beneficial to our classroom, the children's eyes were opened to the difficulty that their peers face every day in other countries. The project itself allowed the children to focus in on several areas that interested them and they enjoyed researching for this during school time and independently on their own. We did not do a whole class project, there were small groups involved and the children had the opportunity to present their work to the class and the school, which increased their confidence and following questions and answers it focused their study and research. It was a project with a difference because all of the facts and statistics were current and relevant to the present day, they enjoyed working together and I believe it opened their eyes to the world around them.

6th class teacher, St John the Apostle NS

As a teacher I thoroughly enjoyed working with the children on their Irish Aid One Work Project. We used the Irish Aid material as a stimulus and starting point for our project. Once the children were familiar with the SDGs, they worked together to decide what form their project would take. At this point, I was able to take on more of an advisory and facilitating role, allowing the children to discuss, organise, brainstorm, debate and problem-solve. The children were highly motivated, researching information at home and sharing it at school. The children wanted to share with others the information they discovered about the SDGs and the developing countries. They wanted to show all the work Irish Aid has done for these countries while also sharing with others why these developing countries are special and unique.

The children decided to spread the message as far as possible. They posted and emailed their booklets to schools all over the world including Australia, America, France, Russia and Dubai. They asked schools to email a photo of themselves with the booklet. The children were truly amazed when they received photos from schools from around the world with their booklet.

The Irish Aid One World Awards can easily be integrated into other subjects like Geography (SDGs, developing countries), English (writing), Maths (calculating distances travelled), ICT (using publisher, Word) and Art (designing projects). It helps develop many skills in children such as working as part of a team, organisation, listening, researching, brainstorming and many more.

Taking part in this award is hugely beneficial for teaching global education. Not only did the children learn a lot about global education, I also learned a lot from the children and the information they discovered.

I would recommend this award as it is very broad allowing the teacher and children huge scope for learning. We took part in the Irish Aid Awards two years in a row. I teach in a multi-grade setting and although the same children took part in the project for two consecutive years, they completed two completely different projects.

Teacher reflection (Eilish Lyons) 3rd–6th class

I really enjoyed taking part in the Irish Aid Awards. It was great when we got photos from schools from around the world with our booklet. Even though we are a small school, we made a difference by spreading the word about the Global Goals for Sustainable Development.

I learned a lot about developing countries and the work Irish Aid is doing to help them. I enjoyed going to Dublin and seeing all the projects other schools did.

I really enjoyed making our *One World, One Future* booklet. We researched different developing countries and shared our information at school. It was great fun putting the booklet together. We were delighted when we made it to the final in Dublin.

Children's reflections

Figure 8.10 Sample of children's work at the Irish Aid One World Awards

Methodological strategies for a pedagogy of hope

Children and young people are creative, passionate and innovative problem-solvers. Framing children as part of the solution rather than part of the problem helps to develop their citizenship, agency and life skills. Education plays a vital role in helping children and young people recognise their contribution and responsibilities as citizens of this global community. Effective global learning initiatives including education about SDGs provide children with skills to make informed decisions and take responsible and age-appropriate actions. Complex environmental and global issues have an impact on current and future generations, therefore children have a right to participate in issues and decisions which affect them. Facilitating an appropriate action component is important for giving expression to children's concerns and initiatives. Otherwise, children may feel powerless in the face of daunting challenges. However, the concept of agency needs to move beyond traditional, liberal ideas of personal choice and individual actions (Valentine, 2011). Caiman and Lundegård (2014: 437), for instance, present agency 'as something that children achieve together in transactions rather than something they possess'. They highlight the value of providing multiple experiences for children to be change agents, where their decisions direct the learning process or action project. Hence, the achievement of 'agency for change' is something that children explore and develop together through direct interactions with their environment.

Agency is not just about thoughts and feelings. It is also about the opportunity and the capacity to act on goals, e.g. teaching children about the importance of recycling and then expecting them to recycle will not automatically change behaviour. Children need to be given the opportunities to change their behaviour. It is only through this experience of taking action that child agency can be developed. Furthermore, children need to experience the power of working collectively. Developing agency is not just about what we teach, it is about *how* we teach. It is not just about developing 'the critical eye', it is about developing 'the hopeful heart'. Developing child agency is not an easy task. It involves cognitive, emotional and strategic dimensions. It has both individual and collective dimensions. In view of this, agency is explained as something that children achieve together in transactions rather than something they possess (Caiman and Lundegård, 2014).

For many schools, the SDG agenda is an effective framework for mapping and capturing some work which is already taking place, such as human rights education, global citizenship education, environmental education, or education about social justice. While the SDGs do not include a dedicated goal or target addressing children's agency, the document states 'children and young women and men are critical agents of change and will find in the new Goals a platform to channel their infinite capacities for activism into the creation of a better world' (United Nations, 2015: Point 51).

The creation of a supportive and sensitive classroom is essential for constructive dialogue. The importance of developing a class contract or agreement which addresses both values and behaviours cannot be overstated. In order for children to express their opinions and participate in classroom discussions about controversial topics, they need to feel safe and not fear retaliation for comments they make during the class discussion. There are a number of stances a teacher can take when facilitating discussions about controversial topics (Table 8.5). A range of methodologies devised by educationalists, development practitioners and non-governmental organisations are regularly used in the classroom for developing critical thinking, promoting participation and achieving transformative

Table 8.5 Stances a teacher can take when facilitating discussions about controversial topics

---

Neutral chairperson approach: Teacher invites all opinions and viewpoints without sharing personal perspective. Children are encouraged to give reasons for their opinions.

Balanced approach: Teacher presents a wide range of views on an issue including conservative and radical perspectives, right-wing and left-wing ideologies, using language and issues which are age appropriate.

Devil's advocate approach: This requires the teacher to consciously take up the opposite position to the one expressed by children. This has the advantage of ensuring a range of viewpoints are expressed.

Official line approach: This requires the teacher to promote the side dictated by public authorities.

---

change (Table 8.6). There is no good reason why controversial issues should be avoided in the classroom and for those who may be anxious about teaching these issues, support is available (see publications at the end of this chapter).

Global learning involves many participative learning methodologies including role play, discussion, debate, ranking exercises, cause and consequence activities, working with images and drama techniques. While these methods are not unique to global learning, when used in conjunction with a global and justice perspective they can advance global understanding, global citizenship and individual agency while fostering skills such as critical thinking, questioning, communication and cooperation. They also enable learners to explore, develop and express their own values and opinions while listening respectfully to others' opinions. Examples of these methodologies are set out in Table 8.6.

Table 8.6 Sample of methodologies for promoting participation and critical thinking in primary classrooms

Walking debate	Walking debate is a useful and fun strategy to develop children's communication and critical thinking skills. Ask the children to stand. Place two signs 'Agree' and 'Disagree' on two opposing walls. Explain that children should decide whether they agree or disagree with each statement and stand close to the sign of their choice. Each child should have a reason for their choice. If they are unsure they can stand in the middle of the room. Once a child agrees with another child he/she can move closer to the speaker. The best walking debates feature a lot of movement as children's minds are changed by powerful arguments made by their peers. Walking debates are extremely useful for giving every child a voice regardless of whether they actually speak out. They also provide an opportunity for children to explore the grey areas of difficult issues and encourage them to confront ambiguity. Sample statements: • All children should donate half of their birthday money to charity. • If the school obtains funding it will be invested in a new library. • The field beside our school will be used for building new houses. • The time for geography in our classroom should be increased to two hours every day.

(*Continued*)

Table 8.6 (Continued)

Ranking exercise	Ladder ranking: Give small groups of participants 6 to 12 statements on separate cards or sticky notes. Ask them to place the statements in vertical order of their importance, with the most important at the top of the 'ladder'. Ask groups with the same statements to compare and explain their results.
	Diamond ranking: Give small groups of participants nine statements written on cards or sticky notes. Ask them to arrange the statements in a diamond shape: the most important statement at the top, two statements of equal but lesser importance in the second row, three statements of moderate importance in the third row, two statements of relatively little importance in the fourth row, and the least important statement at the bottom. Ask groups with the same statements to compare and explain their results.
Think pair share	This enables children to think about their own responses to issues and to listen to responses from one other child. It is a useful way of encouraging less vocal children to share ideas initially in pairs and then in larger groups. It also ensures that everyone's views on an issue are represented.
Silent conversation	Writing (or drawing) and silence are used as tools to slow down thinking and allow for silent reflection. This activity uses an enquiry question, markers and a large sheet of paper. Children work in pairs or threes to have a conversation on their sheet of paper. Children can write at will, but it must be done in silence after a reflection on the enquiry question. This strategy is effective for introverts, and provides a readymade visual record of thought for later.
Café conversation	Understanding different viewpoints is a great way to delve deeply into a topic. Five to ten children are given character sheets. These might include details of gender, age, family status (married, single, how many children, etc.), occupation, education level and significant life events. The group is also given a historical event or similar topic.
	Children can create identity charts in collaboration with each other to determine their character's viewpoint informed by geographical research. This is followed by a 'café conversation' for 10–15 minutes.
Town hall circle	Like a real town meeting, individual children are 'given the floor' and a time limit to express their views.
Jigsaw	Jigsaw is a cooperative learning strategy in which the class is divided into small groups consisting of five to six children. These small groups serve as the children's home base. Each member of the home-base group is assigned to an 'expert' group to engage with one aspect of the geographical topic/issue. After the children meet in their expert group and are familiar with their aspect, they return to their home-base groups to share what they have learned with the other group members. This strategy allows everyone in the class to learn all the content relevant to the subject, as opposed to just the piece they were responsible for. The jigsaw strategy can be implemented during one lesson or across a number of lessons, depending on the depth or complexity of the geographical content or skill. It promotes both self- and peer teaching which requires children to understand the material at a deeper level and engage in discussion, problem-solving, and peer teaching. This methodology can be used for three different purposes: • To help children view an issue from multiple perspectives. • To explore several aspects of an issue. • To help children come up with solutions to a problem, or a class action plan.

Speed-dating   Split the class into two groups:

- One group should form an inner circle (seated or standing) and the other group forms an outer circle. Children should be facing each other.
- Pose a question to the class (e.g. 'Do you think Irish people should have to pay water charges?').
- The pairs facing each other should exchange views for approximately one minute. Then ask the outer circle to rotate clockwise.
- Ask this new pair to discuss the question. Continue the rotation until children have had an opportunity to discuss the question with a wide range of partners.
- During these rotations increase the time available for discussion and encourage children to reflect the views they have heard from others. This encourages them to synthesise ideas and share the opinions of others.
- Debrief the activity: Did your opinion change in any way during the carousel? Did you make stronger arguments as you moved to new partners? Did you pick up any interesting views?
- Select one or two children to summarise the views of the whole class based on what they picked up in the carousel.

This method builds up the children's confidence in discussion techniques as they engage only in short discussions. It also allows children to sample a wide range of views without holding a whole-class discussion. It often results in children's refining and expanding their original ideas or thoughts on an issue. It is useful for controversial issues since it allows children to talk in pairs rather than to the whole class; it allows children to arrive at a whole-class conclusion without a whole-class discussion; it encourages active listening.

Conclusion

Learning about people, places and their interconnections within a globalised world is at the heart of geography. Teachers need to be aware of the dangers of reinforcing a single story in efforts to simplify geographical topics. Providing children with different stories and a variety of perspectives and viewpoints is essential.

In the words of Adichie (2009):

> Stories matter. Many stories matter. Stories have been used to dispossess and to malign. But stories can also be used to empower, and to humanize. Stories can break the dignity of a people. But stories can also repair broken dignity.

Teachers also need to be aware of the stories which children may access from other sources. Teaching children to critically evaluate stories and their sources will help children tackle misconceptions and prejudice. Every story shared with children should be chosen with care in the knowledge that stories shape perceptions. Equally, discussions and activities based on stories should encourage children to consider which perspective is included and excluded and associated reasons.

From a sustainability perspective, the way we are currently living cannot be supported by the Earth's resources. In other words, there is no planet B. The SDGs provide a well-resourced opportunity for teachers to explore sustainability issues in the context of economic development, environmental fragility and human aspirations. Preparing young people to live in an interconnected world with the necessary competencies, knowledge, attitudes and skills is a central concern of global citizenship education. The ultimate aim of global citizenship education is transformative in that learners build knowledge, skills, values and attitudes (Table 8.1) to actively contribute to a more sustainable, just, fair, inclusive and peaceful world both now and in the future. Hence, the process of learning needs to be transformative.

Through education for global citizenship teachers can teach from a charity-based or social justice perspective. Considering the pressing issues confronting our global world such as poverty, climate change, war, terrorism and refugees, discussed in Chapter 1, the requirement for a social justice perspective in our schools has never been greater. These issues transcend geographical borders and affect all of us. Geography informed by global competencies involves 21st-century learning, i.e. global learning including creativity, criticality and collaboration. Case studies discussed in this chapter illustrate powerful geographical learning. Teaching primary geography powerfully provides a unique opportunity for children to develop well-informed opinions, take appropriate actions and share their work with a wider audience.

Exercise 8.1 Reflection for teachers

What impact has globalisation had on you, your school and the children you teach?
When you are teaching global issues is there a 'single story' or 'multiple stories' in your planning and delivery?
List all the global activities in your school and sort them into two categories: A. charity/soft global citizenship and B. critical global learning/social justice.
To what extent do you think it is important for schools to teach and learn about social justice over charity or 'soft' global citizenship?

Further resources

CCEA Teaching Controversial. *Issues at Key Stage 3* http://ccea.org.uk/sites/default/files/docs/curriculum/area_of_learning/CCEA_Co
CDVEC and PDSP. (2012) *Teaching controversial issues in the citizenship classroom: A resource for citizenship education* Dublin: CDVEC www.ubuntu.ie/media/controversial-issues.pdf
Doyle, J. and Milchem, K. (2012) *Developing a forest school in early years provision.* London: Practical Pre-School Books.
Knight, S. ed., (2012) *Forest school for all.* London: Sage.
Knight, S., et al., eds, (2013) *Forest school and outdoor learning in the early years.* London: Sage.
Oxfam. (2015a) *Global citizenship guide for schools.* Available on http://policy-practice.oxfam.org.uk/publications/global-citizenship-guides-620105

Articles published in the Geographical Association's journal *Primary Geography*

Arnold, S. and Whittle, J. (2007) Take responsibility. *Primary Geography, 64,* 12–13.
Bakashaba, C. (2016) Enquiring into the global goals. *Primary Geography, 90,* 16.
Chorekdijan, L. (2017) Critical thinking to promote sustainability. *Primary Geography,* Summer *93,* 24–25.
Hicks, D. (2009) Naturally resourceful: Could your school be a transition school make my future sustainable Part 2. *Primary Geography, 70,* 19–21.

Kempster, E. and Witt, S. (2009) Children's ideas on climate change make my future sustainable Part 1. *Primary Geography*, *70*, 16–18.
Morris, G. and Willy, T. (2007) Eight doorways to sustainability *Primary Geography*, Autumn *64*, 34–35.
Owens, P. (2009) Winds of change: Lessons for campus, curriculum and community make my future sustainable Part 3. *Primary Geography*, *70*, 22–24.
Owens, P. (2011) Why sustainability has a future? *Primary Geography*, *74*, 7–9.
Scott, W. (2013) Lessons from sustainability *Primary Geography*, *80*, 14–15.
Whitfield, L. and Harries, J. (2017) Sustainable living at our special schools. *Primary Geography*, Summer 22–23.
Willy, T. and Hatwood, R. (2017) Start gallery interpreting sustainability. *Primary Geography*, *93*, 5.
Woodhouse, S. (2017) Sustaining school gardens. *Primary Geography*, *93*, 16–17.

Useful websites

The World's Largest Lesson: http://worldslargestlesson.globalgoals.org/
The LEAF Programme (international): www.leaf.global/
The LEAF Programme (Ireland): https://leafireland.org/about/
Irish Forest School Association: https://irishforestschoolassociation.ie/
Forest School Association: www.forestschoolassociation.org/
Global Dimension/The world in your classroom: https://globaldimension.org.uk/
Eco Schools Global: www.ecoschools.global/
The Irish National Teachers' Organisation (INTO) has developed a Global Citizenship School (GCS) programme. A GCS is a school which aims to promote a more just, equitable and sustainable world. The establishment of the GCS initiative aims to encourage and facilitate primary schools in learning about and acting upon global issues. The programme is about *learning*, *acting* and *promoting* global citizenship and helping to develop better global citizens. Founded on the values of Justice, Equality and Sustainability for all, the programme supports schools in learning about and acting upon global issues: www.into.ie/ROI/GlobalCitizenshipSchool/
Mission: Explore Food includes activities that encourage children to learn about growing, harvesting, waste and soil. These practical and outdoor-oriented missions have a strong focus on sustainability. Children are encouraged to think sensibly about resources, what happens to our food and where it ends up. It promotes a holistic approach to food with topics such as growing food, harvesting, cooking, eating, waste and soil. Children are encouraged to make scary soup and keep a poo diary, so that they think about their relationship with food. Interdependence is key as children are encouraged to appreciate that their choices affect people in other places. It also promotes critical thinking. By exploring geographies of inequality, children are encouraged to consider systemic unfairness and envisage fairer alternatives. Mission Explore provides a unique simple formula for exploring local environments in an appealing way for children and teachers alike: www.johnmuirtrust.org/initiatives/mission-explore-food

References

Achebe, C. (2006) *Things fall apart*. London: Penguin (originally published in 1958).
Adichie, C.N. (2009) The danger of a single story. TED Talk. www.ted.com/talks/chimamanda_adichie_the_danger_of_a_single_story/transcript
Andreotti, V. (2006) Soft versus critical global citizenship education. *Development Education in Policy and Practice*, *3*, 83–98.
Bonnett, A. (2008) *What is geography?* London: Sage.

Brissett, N. and Mitter, R. (2017) For function or transformation? A critical discourse analysis of education under the Sustainable Development Goals. *Journal for Critical Education Policy Studies (JCEPS)*, 15, 1.

Caiman, C. and Lundegård, I. (2014) Pre-school children's agency in learning for sustainable development. *Environmental Education Research*, 20(4), 437–459.

Darnton, A. and Kirk, M. (2011) *Finding frames: New ways to engage the public in global poverty, executive summary*. Oxford: Oxfam. Available at http://findingframes.org/report.htm (accessed 18 June 2017).

Department of Education and Skills (2014) *'Education for sustainability': The national strategy on education for sustainable development in Ireland, 2014–2020*. Available at www.education.ie/en/Publications/Education-Reports/National-Strategy-on-Education-for-Sustainable-Development-in-Ireland-2014-2020.pdf#sthash.XfgyNiGv.dpuf

Dolan, A. (2014) *You, me and diversity: Picturebooks for teaching development and intercultural education*. London: Trentham Books and IOE Press.

Dolan, A.M. (2012) Education for sustainability in light of Rio+20: Implications for reforms of the B.Ed. Degree programme in Ireland. *Policy & Practice: A Development Education Review*. Review, Vol. 15, Autumn 2012, 28–48.

Griffiths, H. and Allbut, G. (2011) The danger of the single image. *Primary Geography* Sheffield: Geographical Association, 75, 16–17.

Huckle, J. (2010) ESD and the current crisis of capitalism: Teaching beyond green new deals. *Journal of Education for Sustainable Development*, 4(1), 135–142.

Huckle, J. (2017) Powerful geographical knowledge is critical knowledge underpinned by critical realism. *International Research in Geographical and Environmental Education*, 28(1), 70–84.

Hunt, F. and Cara, O. (2015) *Global learning in England: Baseline analysis of the GLP whole school audit 2013–14*. DERC research paper no. 15. London: UCL Institute of Education.

Inman, S. and Rogers, M. (2006) *Building a sustainable future: Challenges for initial teacher education*. Godalming: WWF-UK/CCCI.

Knight, M.B., Melnicove, M. and O Brien, A. S. (illus) 2002. *Africa Is Not a Country*. Minneapolis, USA: Lerner Publishing Group.

Knight, S. (2013) *Forest school and outdoor learning in the early years*. London: Sage.

Kumar, S. (2005) Spiritual imperative. *Resurgence magazine*, 229. www.resurgence.org/2005/kumarspirit229.htm.

Mansilla, V.B. and Jackson, A. (2011) *Educating for global competence: Preparing our youth to engage the world*. New York: Asia Society.

Oberman, R., O'Shea, F., Hickey, B. and Joyce, C. (2014) *Research investigating the engagement of seven- to nine-year-old children with critical literacy and global citizenship education*. Dublin: Education for a Just World Partnership. www.dcu.ie/sites/default/files/chrce/pdf/GlobalThinkingResearchreportbyRowanOberman.pdf

OECD (2018) *Preparing our youth for an inclusive and sustainable world: The OECD PISA global competence framework*. Available at www.oecd.org/pisa/Handbook-PISA-2018-Global-Competence.pdf

Oxfam (2015) *Global citizenship guide for schools*. Available at http://policy-practice.oxfam.org.uk/publications/global-citizenship-guides-620105

Oxfam (2017) An economy for the 99%: It's time to build a human economy that benefits everyone, not just the privileged few. Available at www.oxfam.org/sites/www.oxfam.org/files/file_attachments/bp-economy-for-99-percent-160117-en.pdf

Peterson, A. and Warwick, P. (2015) *Global learning and education: Key concepts and effective practice*. London: Routledge.

Ramanathan, V., Han, H. and Matlock, T. (2016) Educating children to bend the curve: For a stable climate, sustainable nature and sustainable humanity. In Battro, A.M., Léna, P., Sorondo, M.S. and

Von Braun, J., eds. 2016. *Children and sustainable development: Ecological education in a globalized world.* Springer, 3–16.

Reimers, F., Chopra, V., Chung, C.K., Higdon, J. and O'Donnell, E.B. (2016) *Empowering global citizens: A world course.* North Charleston, SC: CreateSpace Independent Publishing Platform.

Selby, D. (2007) The need for climate change in education. Available at www.bneportal.de/coremedia/generator/pm/en/Issue__001/Downloads/01__Contributions/Selby.pdf

Simpson, J. (2016) A study to investigate, explore and identify successful 'Interventions' to support teachers in a transformative move from charity mentality to a social justice mentality, Global Learning Programme Innovation Fund Research Paper No. 2, London: GLP.

Simpson, J. and Barker, L. (2016) Adapted from Andreotti, V., Barker, L. and De Souza, L.M. (2006) *Critical literacy in global citizenship education.* Centre for the Study of Global and Social Justice and Global Education, Global Education Derby.

Spivak, G. (1990) *The post-colonial critic: Interviews, strategies, dialogues.* New York and London: Routledge.

Spivak, G. (2004) Righting wrongs. *The South Atlantic Quarterly, 103,* 523–581.

Sterling, S. (2012) *The future fit framework: An introductory guide to teaching and learning for sustainability in HE.* York: The Higher Education Academy.

Sterling, S. (2015) Commentary on 'Goal 4: Education'. In ICSU, ISSC (2015): *Review of the sustainable development goals: The science perspective.* Paris: International Council for Science (ICSU). www.icsu.org/publications/reports-and-reviews/review-of-targets-for-the-sustainable-development-goals-the-science-perspective-2015

Trivett, V. (2011) 25. US mega corporations: Where they rank if they were countries. www.businessinsider.com/25-corporations-bigger-tan-countries-2011-6?IR=T (accessed 8 January 2018).

UNESCO (2015) Sustainable development goals (accessed 27 January 2017). http://en.unesco.org/sdgs

Valentine, K. (2011) Accounting for agency. *Children & Society, 25*(5), 347–358.

Van Dijk, T.A. ed. (2011) *Discourse studies: A multidisciplinary introduction.* Switzerland: Sage.

Wade, R. (2006) Identifying the issues. In Inman, S. and Rogers, M. eds., *Building a sustainable future: Challenges for initial teacher education.* Surrey: CCCi LSBU/WWF-UK.

Westheimer, J. and Kahne, J. (2004) What kind of citizen? The politics of educating for democracy. *American Educational Research Journal, 41*(2), 237–269.

Appendix 1

Card-sorting activity for teaching about volcanoes

A *volcanologist* is a geologist and scientist who studies the processes involved in the formation and eruptive activity of volcanoes and their current and historic eruptions. Children are asked to read the following cards and divide them into two groups: the volcanologist cards and the personal testimony cards. These cards can be updated as new volcanoes occur. The points raised in these cards can be used to inform further discussions and activities about volcanoes.

Volcanologist cards

A volcano is a vent or 'chimney' that connects molten rock (magma) from within the Earth's crust to the Earth's surface.	At least 50 eruptions rock the Earth each year.	Eruptions can cause the following: lava flows, hot ash flows, mudslides, avalanches, falling ash, floods.
While volcanoes are usually associated with a cone-shaped mountain or hill, they come in a variety of shapes: shield (flat), composite (tall and thin), cinder cones (circular or oval cones) and lava domes (where dome-shaped deposits of hardened lava have built up around the vent).	When magma reaches the Earth's surface it is called lava. When the lava cools, it forms rock.	The Earth's crust is not one solid piece, but many pieces that fit together like a puzzle. These tectonic plates float on the super-heated magma which makes up most of the Earth's interior. When these plates rub against each other, they create pressure. Magma and gases build up. Finally, these explode through holes or vents in the Earth's surface.
Deep under the Earth's crust is magma – rock that is so hot that it is a liquid. Along with this rock are explosive gases. Magma is the Earth's life blood, churning restlessly beneath the crust, and wherever it can it is trying to burst through. It is at these spots that volcanoes form.	The 'Ring of Fire', a 40,000km horseshoe-shaped area of the Pacific Ocean, holds the majority of all the Earth's volcanoes.	A poisonous ash cloud can race down the slope of a volcano at great speed obliterating everything in its path. In AD 79 Mount Vesuvius erupted, devastating the Italian town of Pompeii. The ash deposits preserved the town and the remains of the people within it. They can still be seen today.

Some volcanoes exist underwater, along the ocean floor and even under ice caps.

Volcanoes form when magma reaches the Earth's surface, causing eruptions of lava and ash.

Known as magma inside the volcano and lava on the outside; as it cools and hardens, it forms mountains over millions of years. These mountains are known as volcanoes.

Volcanoes are a force of beauty and rebirth for the planet. Part of the planet's life cycle, lava creates new lands where life can flourish.

Volcanoes create many types of landform.

The soil near volcanoes is said to be rich and fertile, which is why some people actually set up home on the slopes.

The word 'volcano' comes from the Roman name 'Vulcan', the Roman god of fire.

The eruption of Mount St Helens in 1980 created the largest landslide in recorded history.

The world's largest active volcano is Mauna Loa in Hawaii, standing at 4,169m.

The surface of the Earth is called the crust. The crust is broken into massive pieces called plates. Magma flows beneath the crust. Volcanoes often form along the edges of where these pieces or plates meet.

Volcanoes can also occur away from the boundaries of tectonic plates. They can also occur over 'mantle plumes' – super-hot areas of rock inside the Earth.

Volcanoes are classified as 'active', 'dormant' or 'extinct' depending on the amount of volcanic activity happening. 'Active' means there is regular activity, 'dormant' means there has been recent activity but that it is currently quiet and 'extinct' means it's been so long since the last eruption that it's unlikely to ever erupt again.

Sulphur is a naturally occurring mineral that is found primarily near hot springs and volcanic craters. It has a distinct smell like rotten eggs. The sulphur in volcanic ash is a proven, safe age-old remedy that has been used to treat many skin irritations and infections.

Gems, sparkling rubies, garnets and gold can be found inside volcanoes.

Volcanic ash is full of nutrients from deep within the Earth. It provides a wonderful fertiliser for crops. We also use volcanic ash in foot scrubs and body scrubs because of purification and cleaning qualities.

Volcanoes can cause flooding. In Iceland, when the Eyjafjallajökull Volcano erupted under the ice cap, it melted a significant amount of water and that water had to go somewhere.

The eruption of Eyjafjallajökull Volcano in Iceland has had a very positive effect on tourism. Now the biggest industry in Iceland, approx. 1.7 million tourists visit the country each year, five times the population.

The Lava Centre is Iceland's first and only volcanic and earthquake centre focusing on the geology of Iceland.

You will feel the heat and feel the tremble of the Earth.

It is an interactive, high-tech educational exhibition depicting volcanic activity, earthquakes and the creation of Iceland over millions of years.

Personal testimonies/eye witness cards

I rear poultry for the Christmas markets. The turkeys are kept indoors so my business depends on air conditioning. My business is under threat as the ash is creating problems for my air conditioning.

I am a farmer. I have rounded up my sheep and horses and moved them indoors. I know the volcanic ash can bring toxins which are harmful to animals and humans.

I am a professional golfer but it is impossible to play golf at the moment due to the clouds of fine ash in the air.

My flight was cancelled because a volcano erupted in Iceland.

Based on those advisories, over 300 airports in about two dozen countries, and a correspondingly large airspace, were closed in Europe from 15–21 April 2010.

Our houseyard looked like a moonscape – house, water tanks, plants – all a drab grey. The trees along the bank had snapped and lost most of their foliage … inside the house was a depressing mess. The floors were covered in ash. There was even ash on the ceiling and it was caked along the top of the curtains.

The air grew steadily darker and darker, and at 10.30 am we were in total darkness, just the same as on a very dark night. The darkness was so black and intense that I could not see my hand before my eyes.

A red fiery glare was visible in the sky above the burning mountain.

Huge clouds of ash blocked out the sun.

I can't get used to wearing the mask and goggles. I found the smell of the masks nauseating so I've taken to wearing a folded handkerchief over my nose and mouth like a 'cowboy' in a Western movie.

I saw flames and ash falling like raindrops, rivers of lava and clouds of ash.

Children were crying as their panicked parents ran trying to help them escape.

All the roads east of us are closed from our village and the bridge that crosses the river is in danger of being washed away after the Markarfljót River next to our village became flooded.

The sky was lit up like a massive fireworks display.

I can see streams of flood water coming down the volcano.

The whole town has been swept away and I have lost everything except my life.

The ash was mushrooming out in thick clouds – but there was no noise. The earth had stopped moving. It was 6.15 am. We watched in awe. The clouds began to drift towards us. What a tremendous experience! But now the sky was darkening and black specks of ash were falling on us like light rain. There was also an overpowering smell of sulphur.

Thousands are living in temporary shelters, sports centres, village halls and with relatives or friends. Some return to the danger zone during the day to tend to livestock.

There was no power. Every time we left the house we collected ash on our skin and in our hair.

There were parts of trees all over the road, flattened vegetation everywhere and everything covered in ash and mud.

| Because of the ash people are being advised to wear a mask if they go outside and to keep windows shut. Luckily for us, the gas is still drifting eastwards, so we are not getting a huge amount but there's a lot falling in other places. | There was a deafening bang like a giant explosion. A cloud of steam rose up from the top of the volcano. | The ash was dreadful all day yesterday. It was clearly visible in the air and I could feel and see it on my skin every time I stepped outside. |

Appendix 2

Weather glossary

Compiled by Paula Owens, Geographical Association
(Source, Owens. P. (2018) *Weather Glossary* Primary Geography Sheffield: Geographical Association, available on PG_SUM_2018_EXTRA_WeatherGlossary)

Aerosols Tiny particles that remain floating around in the air.
Air temperature A measure of how hot or cold the air is.
Altitude Height as measured above sea level. Altitude affects the weather, because temperatures decrease with height as the air is less dense and does not hold heat as easily.
Anemometer A device used to measure wind speed.
Anticyclone A large-scale high-pressure system (or 'High') where the atmospheric pressure at the surface of the planet is greater than its surrounding environment. Anticyclones are associated with calm weather. High-pressure systems rotate clockwise in the northern hemisphere and anti-clockwise in the southern hemisphere.
Atmosphere The mixture of gases that surround Earth and which have remained relatively stable for the last 200 million years. The atmosphere is divided into five layers (troposphere, stratosphere, mesosphere, thermosphere and exosphere) with most of the weather and clouds found in the troposphere, the lowest layer.
Atmospheric gases The gases that make up Earth's atmosphere. The main ones are nitrogen (78%) and oxygen (21%). The remaining 1% is made up of 0.9% argon and 0.04% carbon dioxide, plus trace amounts of neon, helium, methane, krypton, hydrogen and water. vapour.
Atmospheric pressure A measure of the 'weight' of air pressing down on Earth's surface. Where air is rising we see lower pressure at Earth's surface and where it is sinking we see higher pressure.
Barometer A device used to measure atmospheric pressure, which also indicates other changes in the weather.
Beaufort scale A measure to classify the strength and intensity of the wind. Download the history and full description of the Beaufort scale at: https://www.metoffice.gov.uk/binaries/content/assets/metofficegovuk/pdf/research/library-and-archive/library/publications/factsheets/factsheet_6-the-beaufort-scale.pdf
Biome A community of living things in a large ecological area that share similar characteristics and climatic influences (e.g. a desert).
Biosphere The part of Earth made up of living organisms (animals and plants), whether in the atmosphere, the ocean or on land.

Black ice A thin coating of ice that forms when supercooled drizzle or rain hits a cold surface or when non-supercooled liquid comes into contact with a surface that is well below 0°C. When black ice forms on roads or paths the colour of the surface beneath it is visible.

Blizzard Occurs when moderate or heavy snow is falling, there are wind speeds of 48kph (30mph) or more and visibility is 200m or less.

Climate The long-term weather patterns of a region or different regions. Climate is measured in terms of average seasonal precipitation (rain or snowfall), maximum and minimum temperatures, hours of sunshine, levels of humidity and the frequency of extreme weather events over a given period (the World Meteorological Organization standard is a 30-year average).

Climate change A large-scale, long-term shift in the global mean and variable climate for an extended period of decades or more.

Cloud Clouds form when tiny drops of water or ice crystals settle on particles (aerosols) in the atmosphere. The droplets are so small (have a diameter of one-hundredth of a millimetre) that each cubic metre of cloud will contain 100 million droplets. There are many different types of cloud (including cumulonimbus, cirrus and altocumulus) and even a Cloud Appreciation Society. View the cloud-spotting guide at: www.metoffice.gov.uk/learning/clouds/cloud-spotting-guide or download the cloud fact file at: https://www.metoffice.gov.uk/binaries/content/assets/metofficegovuk/pdf/research/library-and-archive/library/publications/factsheets/factsheet_1-clouds.pdf

Carbon dioxide (CO_2) This naturally occurring gas found in Earth's atmosphere is also a by-product of human activity (such as burning fossil fuels). CO_2 is the principal anthropogenic (caused by humans) greenhouse gas.

Condensation When water vapour comes into contact with a surface that is at or below the dew point, it turns back into liquid (also known as dew).

Cumulonimbus A heavy, dense cloud that can grow very tall – often with an anvil-shaped plume – and is associated with rain, thunder and lightning. It is the only cloud type to produce hail.

Cyclone A large-scale air mass that rotates inwards around a strong centre of low atmospheric pressure.

Desert Any area, cold or hot, that receives fewer than 250mm of rainfall a year.

Dew point The temperature at which the air, when cooled, will become saturated.

Drizzle Rain that is smaller than 0.5mm in diameter, usually falling at rates of 2mm per day or less.

El Niño A large-scale weather phenomenon associated with unusually warm water. El Niño events occasionally form across much of the tropical eastern and central Pacific Ocean every few years as part of a naturally occurring cycle. Both El Niño and La Niña events are accompanied by major changes in the winds and pressure patterns across the tropical Pacific.

Enhanced global warming Occurs when the greenhouse gases released into the atmosphere from human activity trap more heat, causing global temperatures to rise, which results in rapid climate change.

Equatorial climate Describes a region that experiences hot average yearly temperatures and high monthly precipitation.

Extreme weather events Weather that is unusual, unpredictable, unexpected, unseasonal or severe compared with what has occurred in the past or is found in historical records.

Flash flood This occurs when rain falls and/or snow melts so fast that the underlying ground becomes saturated and the water cannot drain away fast enough.

Flood A huge amount of water, submerging a usually dry area.

Fog Caused by tiny water droplets suspended in the air. Fog is basically a cloud at ground level that reduces visibility to fewer than 1000m.

Freezing rain When rain droplets fall through air with a temperature below 0°C and then freeze (form ice) on impact with the ground.

Frost Occurs when cool air causes water vapour in the air to condense and form droplets on surfaces with a temperature below 0°C. When the moisture freezes into ice crystals, this is known as the 'frost point'.

Global warming While this is a natural process that makes Earth just the right temperature for life (as we know it) to exist, enhanced global warming is the acceleration of this process due to human activity.

Greenhouse effect Most infra-red radiation given off by Earth escapes out into space, which has a cooling effect on the planet. However, some heat is trapped by gases in Earth's atmosphere, resulting in a warming effect across the globe.

Greenhouse gases Gases (e.g. CO_2, methane, nitrous oxide) in the atmosphere that absorb the thermal infra-red radiation emitted by Earth's surface, the troposphere and clouds.

Gulf stream A warm current that originates in the Gulf of Mexico and (together with the North Atlantic Drift) crosses the Atlantic Ocean. It transports heat from low to high latitudes and keeps north-west European winter temperatures higher than they would otherwise be.

Hail A form of precipitation falling as round or irregularly shaped pieces of ice (known as hailstones) that start as small ice particles or frozen raindrops. These particles get caught inside cumulonimbus clouds, circulate and grow bigger until the cloud can no longer support their weight, so they fall to Earth.

Heatwave An extended period of very hot weather relative to the expected conditions of the area at that time of year.

Humidity The amount of water vapour in the air.

Hurricane Tropical storms over the Atlantic and north-east Pacific become known as hurricanes when winds reach 119kph (74mph).

Jet stream These ribbons of very strong winds, found 9–16km above Earth's surface, can reach speeds of 322kph (200mph) and move weather systems around the globe.

La Niña This large-scale weather phenomenon is characterised by colder than usual surface ocean temperatures circulating in the tropical East Pacific.

Lightning A giant spark of electrical energy within or between clouds or between a cloud and the ground.

Met Mark An award from the Royal Meteorological Society and Met Office to recognise excellence in weather teaching.

Microclimate The distinctive climate of a small urban or rural area, such as a garden, park or valley.

Mist A suspension of water droplets in the air resulting in a visibility greater than 1000m.

Monsoon A seasonal change from dry to wet associated with the onset of heavy rains, usually in South East Asia.

Northern Lights Light displays (also known as aurora borealis) produced by the collision of charged solar particles (emitted from the Sun) as they interact with Earth's magnetic field in the north polar region.

Okta grid A grid that is used to estimate cloud cover.
Precipitation Any form of water (liquid or solid) falling from the sky. This includes rain, sleet, snow, hail, drizzle and freezing rain.
Rain A form of precipitation that occurs when the water in the air condenses. Warm air can hold more water than cool air, so when warmer air is cooled the moisture condenses to liquid and it rains.
Rainbow An arc-shaped band of coloured light caused by the refraction, reflection and dispersion of sunlight in water droplets.
Sleet Raindrops that have frozen before they hit the ground (or us!). Usually occurs with freezing rain.
Snow A solid precipitation of tiny ice crystals at temperatures well below 0°C, but as larger snowflakes at temperatures near 0°C.
Snowflake Occurs when tiny droplets of super-cooled water freeze in the sky to create an ice crystal. When the air temperature is near 0°C these ice crystals clump together to form snowflakes. Snowflakes (see www.metoffice.gov.uk/learning/snow/snowflake) have six sides or points because of the way they grow; as they float around in the air, the ice crystals join together in the most efficient way: as hexagonal structures.
Storm A violent disturbance of the atmosphere with strong winds (typically 88–119kph or 55–72mph), thunder and rain.
Sunshine Energy from the Sun (solar radiation) falling on Earth. Levels of sunshine are measured in different ways. Read about them at www.metoffice.gov.uk/guide/weather/observations-guide/how-we-measure-sunshine.

Index

A to Z books 49–50, 157–159, 163, 176
Acer, D. 188
Achebe, Chinua 220, 221
Ackermann, E. 85
Action Aid: Power Down 141
Adams, Ansel 183
Adichie, Chimamanda Ngozi 220–221, 241
advertising 214
aesthetics 21
Africa 5, 100, 101, 220–222
agency 9, 14, 20, 23, 74, 75, 238; arts 208–209; climate change 129, 130; global learning 212; participative learning methodologies 239; playful approaches 85; 'possibility thinking' 199, 207
Agnew, J. 59
Ainsworth, S. 149
Aistear 85–86, 114
analysis 35, 37
Andreotti, V. 218
application 35, 37
architecture 188–191
artefacts 59, 60, 96–103
arts 8, 181–211; bear hunt activity 77, 78; chainsaw art 196–197; climate change 197–198; eco art 183–184; landscape boxes 205–206; place-based learning 184–188; school grounds 200–201; scrapbooking 203–204; urban knitting 201–203
Asia 5
Askins, K. 91
asylum seekers 17
Australian Aboriginals 165–166
awe 22–23, 29, 85

Barlow, A. 37
Barlow, C. 184
Barnaby Bear 40–41
Barnes, J. 181, 182
Beames, S. 72
bear hunt activity 75–78, 80
The Bees' Big Day Out 195–196
being in the world 59

Biesta, G. 5–6, 23
biodiversity 15, 17, 125, 183, 234
biophilia 183
Björk 183
blogs 24, 25, 148, 165
Bonnett, A. 1, 10
Boon, H.J. 126
Boyd, Arthur 183
Brexit 11, 16, 213
bridges 26–27
Bridgnorth 26
Brook, A. 184
Brooks, C. 34
Bruce, T. 84
Bruner, Jerome 23
Bryson, Bill 22
buildings 188–189, 190
Burren National Park 68–69, 104, 105, 200

café conversations 240
Caiman, C. 238
cameras 155, 156, 160, 166
'capes and bays' approach 4
carbon dioxide emissions 126–127; *see also* greenhouse gas emissions
Catling, S. 14–15, 19–20, 28, 71, 98, 113, 114, 121
CGC *see* critical global citizenship
chainsaw art 196–197
change 5, 15, 16, 41–42, 238–239
charity 218, 219, 242
child-centred approaches 73
child labour 96, 214
China 5, 52
Christian Aid 95, 96, 126, 151
Cicalò, E. 149, 150
citizenship 18, 23, 130, 216, 223, 238; *see also* global citizenship
Claregalway 190–194, 207, 208
class mascots 39–41
climate 12, 154; *see also* weather
Climate 4 Classrooms 141
Climate Action Project 131–135

climate change 8, 15, 113, 124–140, 212, 242; arts 197–198; carbon storage 230; climate justice 126, 140; Cloughjordan ecovillage 224, 225; denial 125; education 127–130, 139–140; environmental globalisation 213; fiction and non-fiction literature 143–144; misconceptions about 125–126, 128, 139; mitigation and adaptation 125, 128; newspaper project 48–49; Sustainable Development Goals 17, 234; unusual weather events 121–124; websites 141
Climate Change in the Classroom 141
Cloughjordan ecovillage 224–226
collaboration 1, 18, 19; arts-based learning 208; Climate Action Project 133, 135; eco art 183; enquiry-based learning 51; global learning 212, 242; graphicacy, map work and visual literacy 176; Junior Entrepreneur Programme 50; powerful knowledge 21; preparation for life 20; weather and climate change 139; *see also* group work; teamwork
colonisation 214
communication 18; architecture 189; critical thinking 41–42; geographical thinking 35; global citizenship 217; global competency 215; global learning 212; participative learning methodologies 239; play 84–85, 87; preparation for life 20
concepts 35
Concern 151
conflict 12, 15, 16–17, 217
connections 2, 3, 35
Connemara National Park 184–188
conservation 183, 212
constructive play 86
constructivism 8, 44, 59–60; drama 189; playful approaches 85; powerful knowledge 19; robotics 137
consumerism 13
continents 11
continuing professional development (CPD) 24, 62, 216
controversial issues 19, 42, 216, 238–239, 241
corporations 213
countries 11
Craft, A. 15, 17, 20, 181–182, 207
Creating Futures 141
creativity 1, 18, 19, 50, 181–182; architecture 189; arts-based learning 208; culture kits 99; eco art 183; global learning 212, 242; play 85, 86; preparation for life 20; scrapbooking 204
Cresswell, Tim 59
critical global citizenship (CGC) 218–219
critical global learning 220–222
critical pedagogy of place 74–75
critical thinking 2, 5, 14, 18, 19, 29; arts-based learning 207, 208; Climate Action Project 135; critical geographical thinking 41–42; global citizenship 217, 218–219; global learning 212, 220, 242; graphicacy, map work and visual literacy 176; participative learning methodologies 239; play 89; powerful knowledge 20; weather and climate change 139
Csikszentmihalyi, M. 85, 181
culture: artefacts 98, 99, 103; asking questions 40; cultural globalisation 214; global competency 216; intercultural education 3–4, 61
culture kits 99, 103
curiosity 1, 55, 84, 151; artefacts 96, 98, 99; place-based 60–61, 62, 72; questions 38
Curraghchase Forest 230–232
curriculum: Aistear early childhood education framework 85–86; child-centred approaches 73; curriculum making 28, 35, 37, 54, 97; facts and figures 20; GA Curriculum Proposals 11–12; weather 114
cyclones 119, 122, 124–125

dance 181, 182, 207
Daniels, H. 50–51
data collection 35, 44, 130
data handling skills 18
decision-making 3, 5, 29; arts-based learning 207; 'big ideas' 11; geographical thinking 35, 54; playful approaches 84; public involvement 209
deep learning 54–55, 85
Demarest, A.B. 72, 75
den building 21, 70–71
detectives 99
Dewey, John 23, 44, 65, 182
DiCaprio, Leonardo 197–198
didactic pedagogies 51
digital technology 16; *see also* technology
Dioum, Baba 73
discrimination 222
discussions 239, 241
diversity 17, 41–42, 216, 217, 235
'doing' geography 7, 24
dolls 100
Donnelly, Joanna 119
Dorling, D. 10
drama 181, 182, 189–190, 239; place-based learning 59, 60, 76–77; socio-dramatic play 86
drawing 149–151, 175
drinks 100
droughts 15, 124, 128
Dubai 132–134

early childhood education 85–86
Earth-opoly 94
earthquakes 15
eco art 183–184
Eco Detectives 141

eco-playful pedagogy 129
Eco Schools 226
economic globalisation 16, 213
economics 11
ecosystems 11
ecovillages 224–226
Edelson, D.C. 34–35
education 5–6; climate change 127–130, 139–140; education for all 14; place-based 65–66, 68, 72, 74, 78; Sustainable Development Goals 234, 235
Education for Sustainability (EfS) 223
8-Way Thinking 7, 66–68, 80
Eisner, E.W. 207
emotional literacy 21
empathy 21, 23, 217, 220
empowerment 19–20, 50, 219
energy: GA Curriculum Proposals 11, 12; renewable 128; rising costs of 212; Sustainable Development Goals 234
enquiry 7, 8, 38, 42–44, 54–55; artefacts 103; multiple intelligences 66–67; newspaper project 44–49; place-based education 65; teachers' roles 50–51
enquiry-based learning 2, 23, 42–44, 54–55, 70, 127; artefacts 96; climate change 130; geographical thinking 35; questions 38; teachers' roles 50–51; weather 118
environment 3, 11–12, 35, 36; *see also* sustainability
environmental globalisation 213
environmental knowledge 1
equity 217
erosion 103
ethnogeography 28
Europe 11, 15–16
evaluation 35, 37
everyday items 12–13
experiential learning 2, 4, 8
exploratory play 86
Eyre Square 201–203
Eyrecourt 44–49, 151, 157–159

fabric 100
Facebook 2, 130, 136, 148
famine 15
field sketching 149–151
Field, S.L. 98
field trips 21, 59, 65, 80
fieldwork 12, 35, 62, 73, 114–115
fires 119–120
FIRST LEGO League 135–136, 137
Flat Stanley 78–80
flooding 15, 136–137; arts-based learning 208; Claregalway 191, 193; climate change 124, 128, 129; flood defences 138; mapping wetlands 165, 166

Fogarty, Will 196
followthethings.com 30
food 3, 11, 40
food security 126, 234
Forbus, K.D. 149
forecasting 117–118, 120
forest fires 15
Forest School 232–233
forests 230–232, 234
Fourth Industrial Revolution 16
free movement 16
Freire, Paulo 23, 220
Froebel, Friedrich 73, 182

GA *see* Geographical Association
Galway 93–94, 95, 132, 136, 201–203
games 86, 88, 93–96, 109, 182
Gardner, Howard 7, 23, 24, 66, 182
gender 74, 234, 235
Geo-capabilities Project 19
geo-literacy 165
geo-photo challenges 156–157, 164
Geographical Association (GA) 10, 11–12, 30, 40
geographical knowledge 19–21, 28, 35; Climate Action Project 135; curriculum making 37; photographs 175; *see also* knowledge building
geographical thinking 7, 34–38, 54, 139; critical 41–42
geography, definition of 10–11
geology 103–108
Gilbert, Elizabeth 22
Gilbert, Ian 66–67
Giroux, H. 220
global citizenship 17–18, 215–217, 235, 238, 239, 241–242
global competency 8–9, 18, 212, 214–215, 218, 242; arts-based learning 208; Climate Action Project 135; education for 215–217; graphicacy, map work and visual literacy 176; powerful knowledge 21; weather and climate change 139, 140
Global Learning Project (GLP) 216
Global Schools' Climate Action Project 131–135
global warming *see* climate change
globalisation 1, 8, 212, 213–214, 217; challenges of 15; future of 16; narrative of 5; visual literacy 147
Golden rectangles 188–189
Goldsworthy, Andy 183, 186
Google 2, 16
Google Drive 45
Google Earth 16, 37
Google Maps 148, 164, 172, 175
Google StreetView 61
governance 217
Gozen, G. 188
grammar of geography 35

Granard 49–50
graphicacy 8, 148–149, 172, 175–176
graphical intelligence 147, 149
Green Schools 169–171, 226–228, 229
Greene, M. 87
greenhouse effect 125–126, 141; *see also* climate change
greenhouse gas emissions 124, 139; *see also* carbon dioxide emissions
Greenwave Project 116
Greenwood, D 61
group work 52, 189, 240–241; *see also* collaboration; teamwork
growth mindset 18
Gruenewald, D.A. 58–59, 74
guerrilla geography 91

Harry Potter and the Philosopher's Stone 173
Hart, A. 208
Hart, R.A. 164
Harvey, S. 50–51
hazards 12, 119–121, 124–125
Heaney, Seamus 72
Hemingway, Ernest 22
Hicks, D. 15, 129, 199
Higgins, Michael D. 187–188, 198
Higgins, P. 69
The Hobbit 173
holistic geography 35
hooks, bell 74
Huckle, J. 19
human rights 216, 217, 223, 235, 238
hurricanes 119–120, 121–124, 130, 132–133
Hutchins, Pat 174
Hutson, G. 65

identity 69, 217
ideology 220
igneous rocks 104, 105, 106
images 147–148, 151–163, 175, 176, 203–204, 205, 239; *see also* photographs
imaginative play 86, 87–88, 91
India 5
inequalities 12, 15–16, 216; critical global citizenship 219; globalisation 213; rise in 214; Sustainable Development Goals 212, 234
information 35
innovation 18, 212, 234
Instagram 148
intercultural education 3–4, 61
interests 38
international events 4
international knowledge 1
internet 16, 17
investigations 23, 38, 43, 51, 55
Ireland: architecture 188–189; bridges 26–27; carbon dioxide emissions 126–127; Climate Action Project 132; definition of geography 10; Education for Sustainability 223; flooding 15; Greenwave Project 116; *Keep on Track Project* 24–26; landscape boxes 206; place names 70; *Proclamation for a New Generation* 199–200; rocks 105; trees 230; unusual weather events 122–123; urban landscape 190–191
Irish Aid Our World Awards 235–237
Ironbridge 26
Israel 40–41
Israel, A.L. 65, 74

Jackson, P. 34
jigsaw strategy 240
journey boxes 98
journey sticks 165–168
Junior Entrepreneur Programme (JEP) 49–50
justice-based approach 218–219, 242; *see also* social justice

Kafai, Y.B. 94
Kagawa, F. 125, 127, 128, 140
Kahne, J. 218
Karkdijk, J. 52
Katie Morag 173
Keep on Track Project 24–26
Kent, G. 203
Kenya 5
Kidman, G. 38
knitting 201–203
knowledge building 18, 20; arts-based learning 207; Climate Action Project 133, 135; graphicacy, map work and visual literacy 176; weather and climate change 139; *see also* geographical knowledge
knowledge construction 44, 55
Kramarski, B. 38
Kress, G. 89
Kumar, S. 223

Labbo, L.D. 98
Lambert, D. 10, 35
landforms 154
landmarks 77, 169, 174, 206
landscape boxes 205–206
landscapes 2; creative teaching 182; eco art 183; GA Curriculum Proposals 11, 12; literature 72–73; photographs 154; place 72
Lane, R. 113, 114, 121
language 39, 70; Irish 161, 163; play 86, 109; weather 119
Laye, Camara 220
learning: Aistear early childhood education framework 85–86; artefacts 98; deep 54–55, 85; education for all 14; outdoor 2, 7, 73–74, 231, 232; place-based 58, 59–60,

65, 69, 184–188, 206; play-based 84, 108; scrapbooking 204; transformative 242; *see also* enquiry-based learning
Learning about Forests (LEAF) programme 230–232
Leat, D. 43
Lee, C. 10
Lego 90, 132, 135–138, 189
Leonard, Annie 13
Limerick Smart Travel School Project 168–171
literacy 14, 50, 62, 204
literature 72–73, 142–145, 172–175
'living geography' 10–11
Living Planet Index 15
local places 61, 62–65, 69, 184–188
location 2, 39, 40
Long, Richard 183
Louv, R. 73
Lundegård, I. 238

M-Pesa 5
MacFarlane, R. 70
Mackintosh, M. 147, 203
manipulative play 86
Mapmaker Chronicles 174
maps 164–165, 167, 175–176, 177, 209; creative teaching 182; eco art 183; GA Curriculum Proposals 11; graphicacy 148, 149; landmarks 77; literature 172–175; place-based education 65; trail booklets 63; transport 171–172; travel maps 169
Marshall, T. 16
Martin, F. 14, 28
mascots 39–41
masks 101
materials 13, 88–89
McInerney, P. 65–66, 74
media 2, 11, 19, 148; *see also* social media
memories of learning 4, 28
Met Éireann 119, 123, 132, 140
metacognition 38
metamorphic rocks 104, 106
methodological strategies 238–241
MetLink 141
Mevarech, Z.R. 38
migration 15; *see also* refugees
Millennium Development Goals (MDGs) 17, 233
Milne, A.A. 173
miniature figures 89–91
miniature landscapes 205–206
Mission Explore 91–93, 109
mnemonics 4
mobile devices 2, 16
money 5
Monopoly 93–95
Montessori, Maria 73
Morgan, A. 28

Morgan-Jones, Tom 109
movement 11
Moyles, J. 84
multimodal learning 51, 60
multiple intelligences 7, 66–67, 182, 207
music 59, 60, 181, 182, 195–196, 207
musical instruments 101
mysteries 52–54

names 70
National Primary Science Fair 165
Native Americans 165
natural disasters 15, 119–121, 136
naturalistic intelligence 7, 67
nature 22, 72, 85, 162, 230–232
nested hierarchy 37
New Zealand 86
newspaper project 44–49
non-governmental organisations (NGOs) 126, 151
numeracy 14, 50
Nussbaum, M. 19

Oberman, R. 222
observational skills 150, 156–157
obstacle courses 77
oceans 11, 12
The Once Upon a Time Map Book 173
Oranmore 159–163
Organisation for Economic Co-operation and Development (OECD) 17–18, 38, 212, 214–215, 216
orienteering 21
Our World Awards 235–237
outdoor learning 2, 7, 73–74, 231, 232
Owens, Paula 19, 62, 119
Oxfam 126, 141, 151, 157, 214, 216, 217
ozone layer 125–126

Pacini-Ketchabaw, V. 89
Paper Bag Game 96
Paris Climate Conference 126
Parkinson, Alan 109, 205
participative learning methodologies 239–241
patterns 35
Payne, P.G. 69
PBE *see* place-based education
peace 217, 234, 235
pedagogic content knowledge (PCK) 14–15
pedagogy: climate change education 128–129; constructivist 19, 44; critical pedagogy of place 74–75; didactic 51; eco-playful 129
Perry, J. 38
Pestalozzi, Johann 21
Philosophy for Children (P4C) 66
photographs 11–12, 23, 151–163, 176; arts-based learning 186, 204, 205, 207; geographical knowledge 175

physical play 86
Pickering, S. 71, 182
Pickford, T. 38
Piff, P.K. 22
place 3, 7, 58–83; 8-Way Thinking 66–68; architecture 189; attachment to 21, 61, 69; Burren National Park 68–69; critical pedagogy of 74–75; GA Curriculum Proposals 11–12; geographical thinking 35, 36; identity 69; names 70; outdoor learning 73–74; place-based learning 58, 59–60; place making 70–71; scrapbooking 203–204; trail booklets 62–65
place-based education (PBE) 65–66, 68, 72, 74, 78
place-based learning 65, 69, 80, 184–188, 206
planning 190–194, 207–208
plastic 52–54, 228–230
playful approaches 7, 70, 84–112; artefacts 96–103; creative approaches 182; eco-playful pedagogy 129; games 86, 88, 93–96, 109, 182; imaginative play 87–88; Mission Explore 91–93; rocks 103–108; unstructured materials 88–89
podcasting 25, 148
poetry 72–73, 187–188
political globalisation 213
politics 18–19
pollution 125–126
population growth 5, 212
populations 11
'possibility thinking' 181–182, 199, 207
Pound, L. 109
poverty 126, 213, 242; critical global citizenship 219; Sustainable Development Goals 17, 212, 234
power 59, 217, 218
powerful knowledge 19–21, 139
PowerPoint presentations 13, 79, 150, 157, 160–162
prejudice 74, 219, 222, 241
pretend play 86
problem-based learning 8, 42–43
problem-solving 1, 5, 18, 34, 238; architecture 189; arts-based learning 184, 208; 'big ideas' 11; Climate Action Project 133, 135; collaborative 14; critical thinking 42; flood defences 138; geographical thinking 54; graphicacy, map work and visual literacy 176; jigsaw strategy 240; Junior Entrepreneur Programme 50; playful approaches 84, 85, 89; powerful knowledge 21; preparation for life 20; weather and climate change 139
processes 35, 59–60
Proclamation for a New Generation 199–200

questions 35, 37, 38–39; artefacts 102; arts 203, 207; enquiry 7, 38, 43, 44, 46–47, 54, 66–67; images and photographs 152–154; seasons 115; stereotypes 223; weather and climate change 139

race 74
railways 24–26
rain 115
ranking exercises 239, 240
Rauhala, Osmo 182
Raven-Ellison, Daniel 91, 109
Rawling, E. 58, 59, 60, 72, 80
Rawlinson, S. 67
recycling 52, 53, 238
reflection 217
refugees 16–17, 242
renewable energy 128
replicas 97, 101
resilience 1, 18, 125, 232
resources 11, 12, 15, 17
responsibility 23
rhymes 4
Road Map for 21st Century Geography Education Project 35
Roberts, M. 29, 43
robotics 135–138
Roche, Vivienne 198
rocks 103–108
role play 76–77, 86, 154, 239
Rooney, T. 113
Rory's Story Cubes 88
Rosie's Walk 174–175
Royal Meteorological Society 119, 141
Ruggiero, V.P. 42

safe space 216–217
scaffolding 51
scale 35, 36
The Scary Places Map Book: Seven Terrifying Tours 173
scavenger hunts 157
school grounds 200–201
school travel 168–171
Scoffham, S. 22, 62, 181, 182
scrapbooking 203–204
sculpture 183, 195, 196–197, 198, 201–203
SDGs *see* Sustainable Development Goals
sea level rise 122, 124, 128
seasons 114–117, 144–145
sedimentary rocks 104, 106
Selby, D. 125, 127, 128, 140, 223
self-esteem 87, 232, 233
Sen, A. 19
sense of place 7, 60–61, 69
senses 156–157
Signs of Autumn Twitter Project 115–117
silent conversations 240
simulation games 95–96
sketching 149–151, 175
Skovbjerg, Helle Marie 85
Skylink-Pro 130
Skype 131, 132–133
small world play 86, 89–91

Sobel, D. 65, 72, 164
social class 74
social justice 74, 216, 217, 218–219, 238, 242
social media 11, 19, 148, 165, 176, 222, 235
social networks 17
social skills 87, 232
socio-dramatic play 86
soil 154
soundscapes 195–196
Sourcemap 31
South America 5
space 3, 11–12, 35, 36, 189
spatial thinking 148–149, 176, 177
speed-dating 241
Spivak, G. 214, 220
sports events 4
Stedman, R. 60
Steer, Helen 109
Steiner, Rudolf 73
stereotypes 219, 221, 222–223
Sterling, S. 223
Stone, M.K. 72
stories 59, 60, 71, 241; bear hunt activity 78; critical global learning 220, 221; play 86, 88
storms 119–120, 124–125, 128
The Story of Stuff 13, 30
Street Maps 16
streetscapes 151
sustainability 11, 182, 213, 216, 223–233, 241
sustainable development 215, 217, 235
Sustainable Development Goals (SDGs) 17–18, 212, 233–235, 238, 241; Climate Action Project 133; education for sustainability 223; Irish Aid Our World Awards 235, 236
sustainable travel 168–171
systems-thinking 12

Tait, A.L. 174
Tanner, J. 21, 61
Te Whariki 86
teachers: climate change 128, 139; curriculum making 37, 54; enquiry 50–51; memories of learning 4, 28
teaching 14–15; *see also* pedagogy
teamwork: arts-based learning 208; Climate Action Project 135; Forest School 233; graphicacy, map work and visual literacy 176; Lego robotics 137; powerful knowledge 21; weather and climate change 139; *see also* collaboration; group work
technology 1, 17, 55; Climate Action Project 133; cultural globalisation 214; Fourth Industrial Revolution 16; images 176; Junior Entrepreneur Programme 50; *Keep on Track Project* 24, 25; M-Pesa 5; newspaper project 45; photographs 155; visual literacy 148; *see also* Web 2.0 tools

tectonics 11, 103
teddies 39–41
Telford 26
terrorism 15, 213, 242
Theroux, Paul 22
think pair share 240
Tidy Towns 71
tourism 15, 34, 104, 162–163, 224, 247
Tovey, H. 72
town hall circles 240
toys 97, 100
trade 5, 15, 16, 214
trails 59, 60, 62–65, 80, 206
transport 64, 162, 194, 214
travel 3, 64, 168–171
trees 115, 185–186, 230–232
Trócaire 126, 141, 151
Trump, Donald 16, 125, 213
Tuan, Y.F. 69
21st-century skills 5, 6, 10, 17–18, 19; architecture 189; arts-based learning 207–208; Climate Action Project 133; critical geographical thinking 41–42; decision-making 29; eco art 183; global learning 212, 242; graphicacy, map work and visual literacy 148, 176; preparation for life 20; weather and climate change 139
Twitter 45, 115–116, 130, 136, 148, 165

UNESCO 14
United Nations 17, 235
United States 15–16
urban knitting 201–203

values 217
Van Dijk, T.A. 220
vegetation 154
videocasting 148
visual literacy 8, 147–148, 164, 175–176
vocabulary 35, 39, 77, 119, 174, 177
voice 20, 74, 219
volcanoes 15, 103, 105–106, 246–249
Vygotsky, L. 44, 51, 182

Waldron, F. 129, 130
walks 160–163, 166–167, 183; walkability audits 169–170, 171; walking debates 239
war 15, 242; *see also* conflict
water: GA Curriculum Proposals 11; Green Schools initiative 227–228; Lego 137–138; photographs 154; shortages 15; Sustainable Development Goals 234
Wattchow, B. 69
weather 8, 40, 113–146; fiction and non-fiction literature 142–143; forecasting 117–118, 120; glossary 250–253; seasons 114–117; unusual events and natural disasters 119–124; vocabulary 119; *see also* climate change

weathering 103
Web 2.0 tools 24, 25, 133, 148
WeDo 137
We're Going on a Bear Hunt 75–78
Westheimer, J. 218
wetlands 165, 166
WhatsApp 148
Whitehouse, S. 37
Wiegand, P. 164
wildlife 15
Williamson, C. 208
wind 11, 12, 115; *see also* hurricanes
Winnie the Pooh 173–174

Witt, S. 70–71, 89, 90
wonder 22–23, 29, 85, 182
Woodyer, T. 84
World Bank 15
World Trading Game 95, 96
World Wildlife Fund (WWF) 15
World's Largest Lesson (WLL) 235
writing 240

Young, M. 19
YouTube 20, 105, 148

Zone of Proximal Development (ZPD) 51

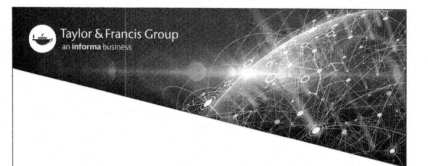